Internal Combustion Engine Fundamentals

Internal Combustion Engine Fundamentals

Zelda Hansen

New York

Published by NY Research Press
118-35 Queens Blvd., Suite 400,
Forest Hills, NY 11375, USA
www.nyresearchpress.com

Internal Combustion Engine Fundamentals
Zelda Hansen

International Standard Book Number: 978-1-64725-426-1 (Hardback)

Cataloging-in-Publication Data

Internal combustion engine fundamentals / Zelda Hansen.
p. cm.
Includes bibliographical references and index.
ISBN 978-1-64725-426-1
1. Internal combustion engines. 2. Gas producers. 3. Engines. 4. Heat-engines. 5. Engineering.
I. Hansen, Zelda.

TJ805 .I58 2023
621.43--dc23

Contents

Preface

An internal combustion engine (IC engine) refers to a type of heat engine wherein the combustion of fuel occurs with the help of an oxidizer in the combustion chamber, which is a significant part of the working fluid circuit. The expansion of the high-pressure and high-temperature gases generated through combustion puts direct force on certain components of an IC engine. Usually, the force is applied to turbine blades, pistons, a nozzle, or a rotor. The component is moved across a distance by this force, which converts chemical energy into kinetic energy, which is further utilized to propel, power or move whatsoever the engine is coupled with. This book is compiled in such a manner, that it will provide an in-depth knowledge about the theory and working of the internal combustion engine. The various advancements in these engines are glanced at and their applications as well as ramifications are looked at in detail. Those in search of information to further their knowledge will be greatly assisted by this book.

Significant researches are present in this book. Intensive efforts have been employed by authors to make this book an outstanding discourse. This book contains the enlightening chapters which have been written on the basis of significant researches done by the experts.

Finally, I would also like to thank all the members involved in this book for being a team and meeting all the deadlines for the submission of their respective works. I would also like to thank my friends and family for being supportive in my efforts.

Zelda Hansen

1

Thermodynamic Analysis of Cycles, Fuels and Carburetion

1.1 Introduction: Classification

There are different types of IC engines and they can be classified on the following basis:

Thermodynamic Cycle

1. Constant Volume or Otto Cycle-the heat energy is added to the system at constant volume.
2. Constant Pressure or Diesel Cycle-the heat energy is added to the system at constant pressure. Limited Pressure or Dual cycle the heat energy is added to the system partly at constant volume and partly at constant pressure.
3. Joule or Brayton Cycle-heat energy is added at constant pressure and the heat energy is rejected also at constant pressure whereas in Otto, Diesel and Dual cycles, the heat energy is rejected at constant volume.

Number of Strokes Per Cycle

1. 4-stroke engines the thermodynamic cycle is completed in four strokes of the piston.
2. 2-stroke engines the thermodynamic cycle is completed in two strokes of piston.

Ignition System

1. Spark Ignition (SI) a combustible homogeneous air fuel mixture prepared by the carburetor and sucked inside the cylinder is ignited with the help of a spark plug.
2. Compression Ignition (CI) the fuel is injected inside the system and it automatically ignites in presence of high temperature and high pressure air (compressed within the system prior to the injection of fuel) present inside the system.
3. Pilot injection of fuel oil in gas engines.

Fuel Used

1. Petrol engine the engine uses petrol or gasoline as the source of energy.
2. Oil engine uses diesel oil, mineral oil, as fuel.
3. Gas engines use gaseous fuel, coal gas, natural gas, coke oven gas, producer gas etc.
4. Multi fuel engines use gasoline or diesel oil for starting purpose and kerosene or biogas as primary fuel.

Cooling Arrangement

1. Water cooled cylinder walls are cooled by circulating water.
2. Air cooled walls of the system are cooled by blowing atmospheric air over hot surfaces (motor cycles. Scooters, air crafts have air cooled engines).

Arrangement of Cylinders

Multi-cylinders engines are invariably used in automobiles. The upper limit on the size of the cylinder is determined by the dynamic considerations, i.e., when the reciprocating masses of the piston and connecting rod are accelerated and decelerated, inertia forces develop and they put a limit on the speed of the engine and thus on the power output.

Therefore, the displaced volume is distributed amongst several smaller cylinders. By this means, the Inertia forces per cylinder are reduced and the forces in one cylinder can easily be balanced by an appropriate arrangement of other cylinders.

1.1.1 Engine Nomenclature

The major components of the internal combustion engine are:

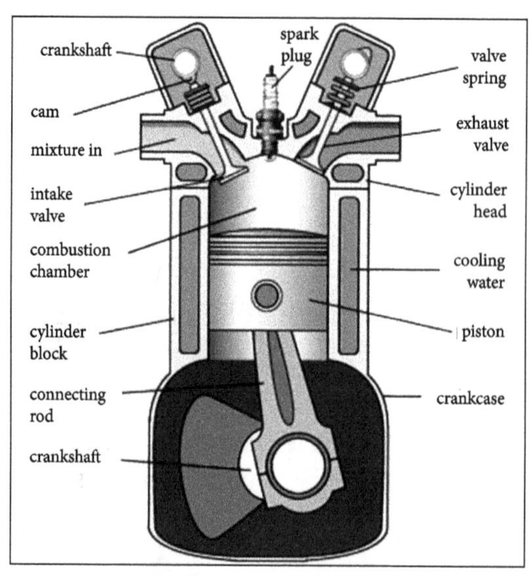

Engine nomenclature.

1. Combustion chambers.
2. Ignition systems.
3. Spark.
4. Compression.
5. Ignition timing.
6. Fuel systems:
   i. Carburettor.
   ii. Fuel injection.
   iii. Fuel pump.
7 Oxidiser-Air inlet system:
   i. Naturally aspirated engines.
   ii. Superchargers and turbochargers.
   iii. Liquids.
8. Parts.
9. Valves:
   i. Piston engine valves.
   ii. Control valves.
10. Exhaust systems.
11. Cooling systems.
12. Piston.
13. Propelling nozzle.
14. Crankshaft.
15. Flywheels.
16. Starter systems.
17. Heat shielding systems.
18. Lubrication systems.
19. Control systems.
20. Diagnostic systems.

Function of IC Engine

1. Cylinder Block

In the bore of cylinder the fresh charge of air-fuel mixture is ignited, compressed by piston and expanded to give power to piston.

2. Cylinder Head

It carries inlet and exhaust valve. Fresh charge is admitted through inlet valve and burnt gases are exhausted from exhaust valve. In case of petrol engine, a spark plug and in case of diesel engine, an injector is also mounted on cylinder head.

3. Piston

During suction stroke, it sucks the fresh charge of air-fuel mixture through inlet valve and compresses during the compression stroke inside the cylinder. This way piston receives power from the expanding gases after ignition in cylinder. Also forces the burn exhaust gases out of the cylinder through exhaust valve.

4. Piston Rings

It prevents the compressed charge of fuel-air mixture from leaking to the other side of the piston. Oil rings, is used for removing lubricating oil from the cylinder after lubrication. This ring prevents the excess oil to mix with charge.

5. Connecting Rod

It changes the reciprocating motion of piston into rotary motion at crankshaft. This way of connecting rod transmits the power produced at piston to crankshaft.

6. Gudgeon Pin

Connects the piston with small end of connecting rod.

7. Crank Pin

Hand over the power and motion to the crankshaft which has come from piston through connecting rod.

8. Crank Shaft

Receives oscillating motion from connecting rod and gives a rotary motion to the main shaft. It also drives the camshafts which actuate the valves of the engine.

9. Cam Shaft

It takes driving force from crankshaft through gear train or chain and operates the inlet valve as well as exhaust valve with the help of cam followers, push rod and rocker arms.

10. Inlet Valve and Exhaust Valve

Inlet valve allow the fresh charge of air-fuel mixture to enter the cylinder bore. Exhaust valve permits the burnt gases to escape from the cylinder bore at proper timing.

11. Governor

It controls the speed of engine at a different load by regulating fuel supply in diesel engine. In petrol engine, supplying the mixture of air-petrol and controlling the speed at various load condition.

12. Carburettor

It converts petrol in fine spray and mixes with air in proper ratio as per requirement of the engine.

13. Fuel Pump

This device supplies the petrol to the carburetor sucking from the fuel tank.

14. Spark Plug

This device is used in petrol engine only and ignites the charge of fuel for combustion.

15. Fuel Injector

This device is used in diesel engine only and delivers fuel in fine spray under pressure.

1.1.2 Engine Operating and Performance Parameters

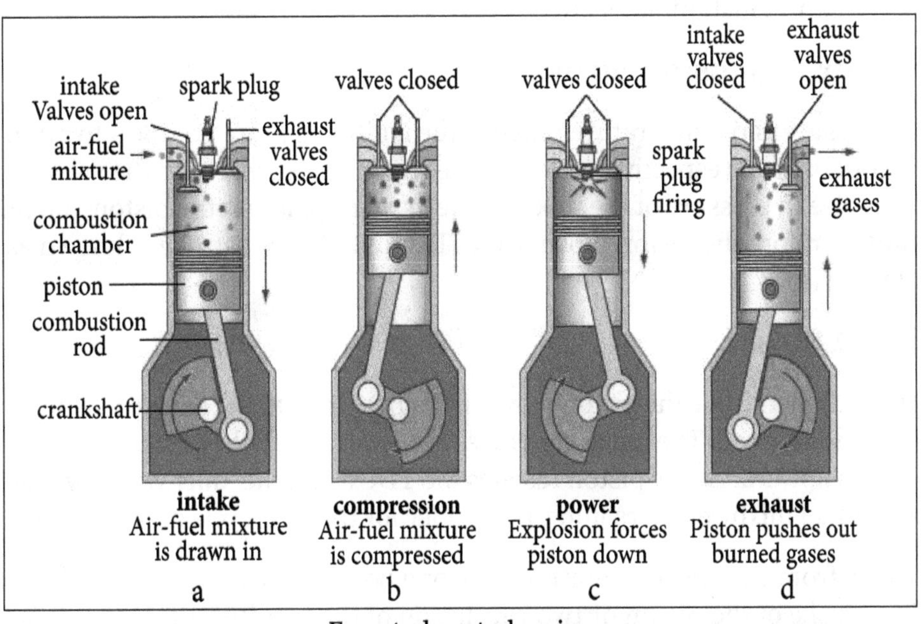

Four stroke petrol engine.

It consists of the following four strokes:

1. Suction stroke.
2. Compression stroke.
3. Power or Expansion stroke.
4. Exhaust stroke.

1. Suction Stroke

At the beginning of the stroke the piston is at the top most position Top Dead Centre and is ready to move downward. As the piston moves downwards vacuum will create inside the cylinder. Due to this vacuum air fuel mixture from the carburetor is sucked into the cylinder through inlet valves till the piston reaches bottom most position Bottom Dead Centre. During the suction stroke, exhaust valve remains in closed condition and inlet valve remains open. At the end of the suction stroke the inlet valve will be closed. Shown in figure (a).

2. Compression Stroke

During the compression stroke both the inlet and exhaust values are in closed condition and the piston moves upward from BDC to compress the air fuel mixture. This process will continue till the piston reaches TDC as shown in figure (b). The compression ratio of engine varies from 6 to 8. The pressure at the end of compression is about 600 to 1200 Kg/m^2.

The temperature at the end of the compression is 250 to 300°C. At the end of this stroke the mixture is ignited by spark plug.it leads to increase in pressure and temperature of the mixture instantaneously.

3. Power or Expansion Stroke

Both the pressure and temperature range of the ignited mixture are 1800 to 2000°C and 3000 to 4000 kN/m^2 respectively. During the expansion stroke both the valves remains closed. The rise in pressure of the mixture exerts an impulse on the piston and pushes it downward therefore the piston move from TDC to BDC. This stroke is known as power stroke, Figure (a).

4. Exhaust Stroke

During the exhaust stroke the piston moves from BDC to TDC the exhaust value is opened and inlet value is closed. The burnt gases are released through the exhaust value when the piston moves upward. As the piston reaches the TDC again the, inlet valves will open and the fresh air fuel mixture enters into the cylinder for the next cycle of operation.

It is obvious from the above operation only one power stroke is produced in both and every four stroke of the piston or two revolution of the crank-shaft. Hence it is termed as four stroke engine.

The performance of the engine is judged from the various parameters listed below:

1. Indicated Horse Power, IHP.
2. Brake Horse Power.
3. Relative fuel Air Ratio.
4. Mechanical Efficiency.
5. Frictional Horse Power.
6. Torque.
7. Fuel Air-Ratio.
8. Thermal Efficiency.
9. Engine exhaust.
10. Mean Effective Pressure.
11. Volumetric Efficiency.
12. Brake Specific Fuel Consumption.
13. Heat Balance.

1. Indicated Horse Power

An IC engine produces power by the combustion of fuel. Power is rate of doing work. Power is the product of torque and angular velocity. The torque is measured by a dynamometer and Rpm by a Tachometer. Power produced by the engine is called Indicated horse power (IHP) and the NET output useful power available from the engine is called the brake horse power (BHP).

$$IHP = p\,L\,A\,N / 60$$

Where,

p=Pressure on the piston, N/m^2.

L=Length of stroke, m.

A=Area of the piston, m^2.

N=Rpm.

IHP-Indicated horse power in Watts.

(i) Torque, T

T = Tangential force × Radius.

(ii) Brake Horse Power

It is useful power available from engine shaft.

$$BHP = 2NT / 60$$

Where,

T-Torque in N.m.

N-RPM.

BHP - Brake horse power in Watts.

(iii) Mechanical Efficiency, η_{mech}

$$\eta_{mech} = BHP / IHP$$

$$\eta_{mech} = \frac{\text{Brake Power (KW)}}{\text{Indicated Power (KW)}}$$

(iv) Frictional Power, Fp

It is the power lost in friction.

$$FP = IHP - BHP$$

(v) Fuel Air Ratio

Fuel-Air Ratio = Mass of fuel / Mass of air.

Fuel air ratio is of three types:

Normal fuel air ratio (called as Stoichiometric fuel air ratio) — Fuel and air are in exact ratio required for complete combustion.

Rich fuel air ratio—Fuel is more and air is less than required for complete combustion.

Lean fuel air ratio—Fuel is less and air is much more than required for complete combustion.

(vi) Relative Fuel Air Ratio

However the relative fuel air ratio is used in the analysis.

Relative fuel air ratio=Actual fuel air ratio/Stoichiometric fuel air ratio.

The performance of an engine is usually studied by heat balance-sheet. The main components of the heat balance are:

1. Heat equivalent to the effective (brake) work of the engine.
2. Heat rejected to the cooling medium.
3. Heat carried away from the engine with the exhaust gases.
4. Unaccounted losses. The unaccounted losses include the radiation losses from the various parts of the engine and heat lost due to incomplete combustion.

The friction loss is not shown as a separate item to the heat balance-sheet as the friction loss ultimately reappears as heat in cooling water, exhaust and radiation.

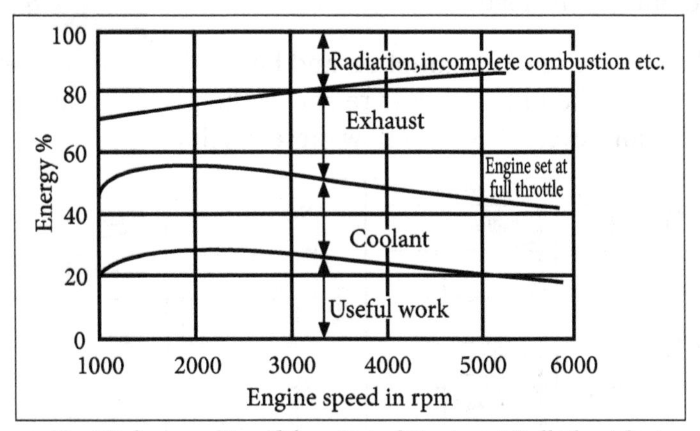

Heat Balance vs Speed for a Petrol Engine at Full Throttle.

The performance of SI engine at constant speed and variable loads is different from the performance at full throttle and variable speed. Figure shows the heat balance of SI engine at constant speed and variable load. The load is varied by altering the throttle and the speed is kept constant by resetting the dynamometer.

Closing the throttle reduces the pressure inside the cylinders but the temperature is affected very little because the air/fuel ratio is substantially constant and the gas temperatures throughout the cycle are high. This results in high loss to coolant at low engine load. This is reason of poor part load thermal efficiency of the SI engine compared with the CI engine.

1. At low loads the efficiency is about 10 percent rising to about 25 percent at full load.
2. The loss to coolant is about 60 percent at low loads and 30 percent at full load.
3. The exhaust temperature rises very slowly with load and as mass flow rate of exhaust gas is reduced because the mass flow rate of fuel into the engine is reduced, the percentage loss to exhaust remains nearly constant (about 21% at low loads to 24% at full load).

4. Percentage loss to radiation increases from about 7% at loads or 20% at full load.

The Performance of a CI Engine

The performance of a CI engine at constant speed variable load is shown in figure:

1. As the efficiency of the CI engine is more than the SI engine, the total losses are less. The coolant loss is more at low loads and radiation, etc. losses are more at high loads.

2. The bmep, bhp and torque directly increase with load, as shown in Figure. Unlike the SI engine bhp and bmep are continuously raising curves and are limited only by the load.

The lowest brake specific fuel consumption and hence the maximum efficiency occurs at about 80 percent of the full load. Figure shows the performance curves of variable speed GM 7850 cc. four cycle V-6 Toro-flow diesel engine.

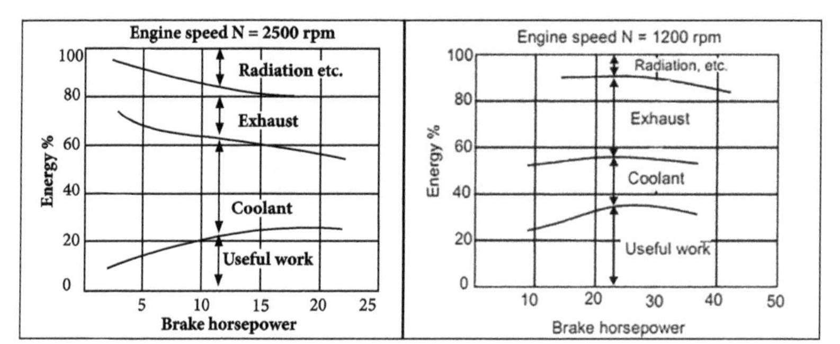

Heat Balance vs. Load for a Petrol Engine Heat Balance vs. Load for a CI Engine.

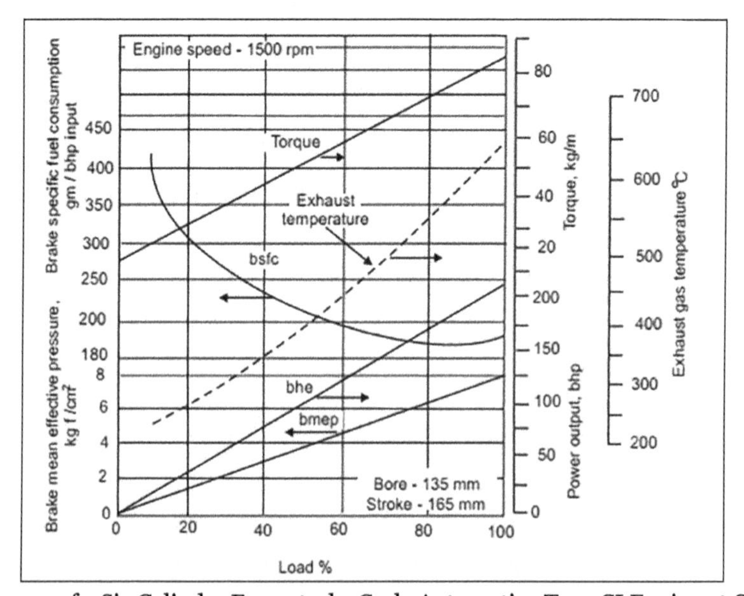

Performance Curves of a Six Cylinder Four-stroke Cycle Automotive Type CI Engine at Constant Speed.

Problems

1. A gasoline engine works on Otto cycle. It consumes 8 litres of gasoline per hour and develops power at the rate of 25 kW. The specific gravity of gasoline is 0.8 and its calorific value is 44000 kJ/kg. Let us determine the indicated thermal efficiency of the engine.

Solution:

Given:

$$m = 8$$
$$P = 25 kW$$
$$g = 0.8$$
$$C_v = 44000 \text{ kJ} / \text{kg}$$

Heat liberated at the input.

$$= m C_v$$

$$= 8 \times \frac{0.8}{60 \times 60}$$

$$= \frac{6.4}{3600}$$

$$\text{Power at the input} = \frac{6.4}{3600} \text{x} \, 44000 \, \text{kW}$$

$$\eta_{ith} = \frac{\text{Output power}}{\text{Input power}}$$

$$= \frac{25}{\frac{6.4 \times 44000}{3600}}$$

$$= \frac{25 \times 3600}{6.4 \times 44000} = 0.3196$$

or, 31.96%

2. A single cylinder engine operating at 2000 rpm develops a torque of 8 N-m. The indicated power of the engine is 2.0 kW. Let us determine the loss due to friction as the percentage of brake power.

Solution:

Given:

$$N = 2000$$
$$T = 8N - m$$
$$P = 2.0KW$$

$$\text{Brake power} = \frac{2\pi NT}{60000} = \frac{2 \times \pi \times 2000 \times 8}{60000}$$

$$= 1.6746 \text{kW}$$

$$\text{Friction power} = 2.0 - 1.6746$$
$$= 0.3253$$
$$\% \text{loss} = \frac{0.3253}{2} \times 100$$
$$\% \text{loss} = 16.2667\%$$

3. A six-cylinder, gasoline engine operates on the four-stroke cycle. The bore of each cylinder is 80 mm and the stroke is 100 mm. The clearance volume per cylinder is 70 cc. At the speed of 4100 rpm, the fuel consumption is 5.5 gm/sec. [or 19.8 kg/hr.) and the torque developed is 160 Nm.

Let us calculate:

i. Brake power,

ii. The brake mean effective pressure,

iii. Brake thermal efficiency if the calorific value of the fuel is 44000 kJ/kg and,

iv. The relative efficiency on a brake power basis assuming the engine works on the constant volume cycle r=1.4 for air.

Solution:

Given:

$$r = 1.4 \text{ for air}$$
$$N = 4100$$
$$T = 160$$
$$m_f - 19.8 \text{kg} / \text{hr}$$

$$bp = \frac{2\pi NT}{60000} = \frac{2 \times \pi \times 4100 \times 160}{60000} = 68.66$$

$$P_{bm} = \frac{bp \times 6000}{LAnK}$$

$$= \frac{68.66 \times 60000}{0.1 \times \frac{\pi}{4} \times (0.08)^2 \times \frac{4100}{2} \times 6}$$

$$= 6.66 \times 10^5 \, P_a$$

$$Pb_m = 6.66 \text{ bar}$$

$$\eta_{bth} = \frac{bp}{m_f \times C_v} = \frac{68.66 \times 3600}{19.8 \times 43000} \times 100 = 29.03\%$$

$$T = 160$$

$$m_f = 19.8 \text{ kg/hr}$$

$$C_v = 43000$$

Compression ratio, $r = \frac{V_s + V_d}{V_d}$

$$V_s = \frac{\pi}{4} D^2 L = \frac{\pi}{4} \times 8^2 \times 10 = 502.65 \text{cc}$$

$$r = \frac{502.65 + 70}{70}$$

$$r = 8.18$$

Air-standard efficiency, $\eta_{otto} = 1 - \frac{1}{(8.18)^{0.4}} = 1 - \frac{1}{2.3179} = 0.56858$

Relative efficiency, $\eta_{rel} = \frac{0.2903}{0.568} \times 100 = 51.109\%$

$$\eta_{bth} = \frac{bp}{m_f \times C_v}$$

$$= \frac{119.82 \times 60}{\frac{4.4}{10} \times 44000} \times 100$$

$$\eta_{bth} = 37.134\%$$

Volume flow rate of air at intake condition

$$a = \frac{6\times287\times300}{1\times10^5} = 5.17\,m^3/min$$

Swept volume per minute,

$$V_s = \frac{\pi}{4}D^2\,LnK$$

$$= \frac{\pi}{4}\times(0.1)^2\times0.9\times\frac{4500}{2}\times9$$

$$= 127.17\,m^3/min$$

Volumetric efficiency, $\eta_v = \frac{5.17}{127.17}\times100$

$$\eta_v = 4.654\%$$

Air-fuel ratio, $\frac{A}{F} = \frac{6.0}{0.44} = 13.64$

4. A six-cylinder, four-stroke gasoline engine having a bore of 90 mm and stroke of 100 mm has a compression ratio 8. The relative efficiency is 60%. When the indicated specific fuel consumption is 3009 g/kWh.

Let us Estimate the,

i. The calorific value of the fuel and,

ii. Corresponding fuel consumption given that i mep is 8.5 bar and speed is 2500 rpm.

Solution:

Given:

r=8

Eff =60%

Specific fuel consumption is 3009 g/kWh.

Rpm=2500

Air-standard efficiency $= 1-\frac{1}{r^{r-1}} = 1-\frac{1}{8^{0.4}} = 0.5647$

$$\text{Relative efficiency} = \frac{\text{Thermal efficiency}}{\text{Air-standard efficiency}}$$

$$\text{Indicated thermal efficiency} = 0.6 \times 0.5647 = 0.3388$$

$$\eta_{ith} = \frac{1}{i_{sfc} \times C_v}$$

$$C_v = \frac{1}{\eta_{ith} \times i_{sfc}} = \frac{3600}{0.3 \times 0.3388}$$

$$C_v = 35417.035 \text{ kJ/kg}$$

$$ip = \frac{P_{im}\, LAnK}{60000}$$

$$= \frac{8.5 \times 10^5 \times 0.1 \times \frac{\pi}{4} \times 0.09^2 \times \frac{2500}{2} \times 6}{60000} = 67.6\,\text{kW}$$

$$\text{Fuel consumption} = \text{isfc} \times ip = 0.3 \times 67.6$$

$$ip = 20.28\,\text{kg/h}.$$

1.2 Valve Timing Diagram of SI and CI Engines

CI Engine

The exact moment at which each of the valves open and close with reference to the position of piston and crank can be shown graphically in a diagram. This diagram is known as "Valve timing diagram".

Theoretical Valve Timing Diagram

In theoretical valve timing diagram, inlet and exhaust valves open and close at both the dead centers. Similarly, all the processes are sharply completed at the TDC or BDC. Figure shows theoretical valve timing diagram for four strokes S.I. Engines:

IVO-Inlet Valve Open.

IVC-Inlet Valve Close.

IS-Ignition Starts.

EVO-Exhaust Valve Open.

EVC-Exhaust Valve Close.

TDC-Top Dead Center.

BDC-Bottom Dead Center.

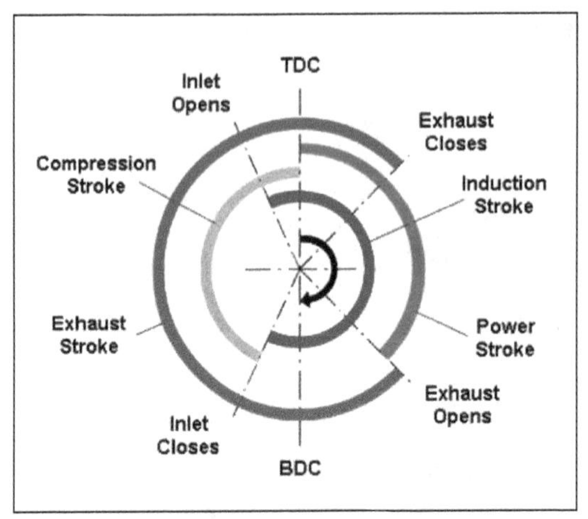

Valve timing Chart.

Actual Valve Timing Diagram

Figure shows actual valve timing diagram for four stroke S.I. engine. The inlet valve opens 10-30° before the TDC. The air-fuel mixture is sucked into the cylinder till the inlet Valve closes. The inlet valve closes 30-40° or even 60° after the BDC. The charge is compressed till the spark occurs. The spark is produced 20-40° before the TDC. It gives sufficient time for the fuel to burn. The pressure and temperature increase. The burnt gases are expanded till the exhaust valve opens.

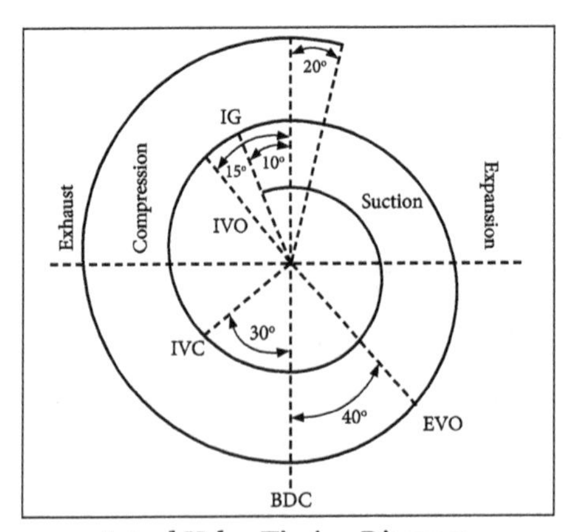

Actual Valve Timing Diagram.

The exhaust valve opens 30-60° before the BDC. The exhaust gases are forced out from the cylinder till the exhaust valve closes. The exhaust valve closes 8-20° after the TDC. Before closing, again the inlet valve opens 10-30° before the TDC. The period between the IVO and EVC is known as valve overlap period. The angle between the inlet valve opening and exhaust valve closing is known as angle of valve overlap.

1.2.1 Comparison of SI and CI engine

S.no.	SI Engine	Cl Engine
(i)	Basic Cycle: SI engine is operated by Otto cycle or constant volume cycle.	CI engine is operated by Diesel cycle or constant pressure cycle.
(ii)	Fuel Used: Fuel used for SI engine is petrol.	Fuel used for CI engine is diesel.
(iii)	Introduction of Fuel: In SI engine, the fuel is introduced to the cylinder along with air through the Inlet valve- during the suction stroke.	In Cl engine, the fuel is injected by the injector at the end of compression stroke.
(iv)	Ignition: In SI engines, the fuel- air mixture is ignited by a high-tension spark plug. Hence it is called as spark ignition engines.	In CI engines, the ignition of fuel air mixture takes place due to the high pressure and temperature of the air. Hence, they are known compression as Ignition Engines.
(v)	Compression Ratio: Compression ratio for SI engine varies from 6 to 8.	Compression ratio for CI engine varies from 12 to 18.
(vi)	Speed: These are used for high speed applications.	These are used for low speed operations.
(vii)	Efficiency: Thermal Efficiency is low for SI engines.	Thermal efficiency is considerably more for CI engines.
(viii)	Weight: Weight is Considerably less for SI engines because of its fewer parts.	Weight is more for CI engines because of more number of parts.

1.3 Modern Developments in IC Engines

- Many different styles of internal combustion engines were built and tested during the second half of the 19th century.
- The first fairly practical engine was invented by J.J.E. Lenoir which appeared on the scene about 1860. During the next decade, several hundred of these engines were built with power up to about 4.5 kW end mechanical efficiency up to 5%.

- The Otto-Langen engine with efficiency improved to about 11% was first introduced in 1867 and several thousands of these were produced during the next decade. This was a type of atmospheric engine with the power stroke propelled by atmospheric pressure acting against a vacuum.
- Although many people were working on four-stroke cycle design, Otto was given credit when his prototype engine was built in 1876.
- In the 1880s, the internal combustion engines first appeared in automobiles. Also in this decade the two-stroke cycle engine became practical and was manufactured in large number.
- Rudolf Diesel, by 1892, had perfected his compression ignition engine into basically the same diesel engine known today. This was after years of development work which included the use of solid fuel in his early experimental engines.
- Early compression engines were noisy, large, slow, single cylinder engines. They were, however, generally more efficient than spark ignition engines.
- It wasn't until the 1920s that multi-cylinder compression ignition engines were made small enough to be used with automobile and trucks.
- Wakdle's first rotary engine was tested at NSV, Germany in 1957.
- The practical Stirling engines in small number are being produced since 1965.
- These engines require costly material and advanced technology for manufacture.
- The advantages of Stirling engine are low exhaust emission and multi-fuel capability.
- Thermal efficiencies higher than 30% have been obtained.

1.3.1 EGR

In internal combustion engines, (EGR) exhaust gas recirculation is a nitrogen oxide (NOx) emissions reduction technique used in gasoline/petrol and diesel engines. EGR works by recirculating a portion of an engine's exhaust gas back to the engine cylinders. This dilutes the O2 in the incoming air stream and provides gases inlet to combustion to act as absorbents of combustion heat to reduce peak in-cylinder temperatures. The NOx is produced in a narrow band of high cylinder temperatures and pressures.

In a gasoline engine, this inert exhaust displaces the amount of combustible matter in the cylinder. In a diesel engine, the exhaust gas replaces some of the excess oxygen in the pre-combustion mixture. Because NOx forms primarily when a mixture of nitrogen and oxygen is subjected to maximum temperature, the lower combustion chamber temperatures caused by EGR reduces the amount of NOx the combustion generates.

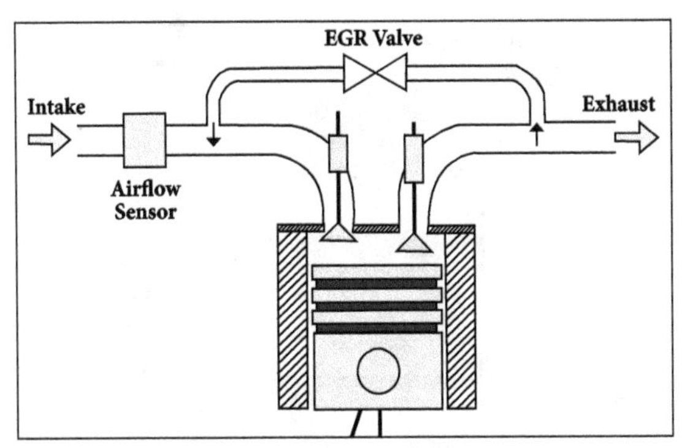

Exhaust gas recirculation.

Gasses re-introduced from EGR systems will also contain near equilibrium concentrations of NOx and CO. The small fraction initially within the combustion chamber inhibits the total net production of these and other pollutants when sampled on a time average. Most of modern engines now require exhaust gas recirculation to meet emissions standards.

Reduced heat rejection: Lowered peak combustion temperatures not only reduce NOx formation, it also reduces the loss of thermal energy to combustion chamber surfaces, leaving more available for conversion to mechanical work during the expansion stroke.

Reduced throttling losses: The addition of inert exhaust gas into the intake system means that for a given power output, the throttle plate should be opened further, resulting in increased inlet manifold pressure and reduced throttling losses.

Reduced chemical dissociation: The lower peak temperatures result in more of the released energy remaining as sensible energy near TDC (Top Dead-Center), rather than being bound up in the dissociation of combustion products.

1.3.2 MPFI

MPFI is a fuel injection technique used in gasoline engines. Multi-port fuel injection injects fuel into the intake ports of each cylinder's intake valve, rather than at a central point within an intake manifold like in spark plugs. It can be sequential, in which injection is timed to coincide with each cylinder's intake stroke.

A separate injector supplies the correct quantity of fuel to each of the engine cylinders by a fuel-rail according to the firing order or in a 'particular sequence'. This system provides further precision by varying the fuel quantity and injection timing by governing the each injector separately and thereby improving the performance and controlling the emissions.

The return valve returns fuel in case the fuel is oversupplied. Also the pressure regulator regulates the pressure of the intake fuel. Fuel filter contains small sized membranes which filters and absorbs the undesirable matters of size 30 to 40 microns.

The fuel and air are mixed in intake manifold and each manifold is controlled by an ECU(Electronic Control Unit), fuel pressure runs between 3 to 5 bars.

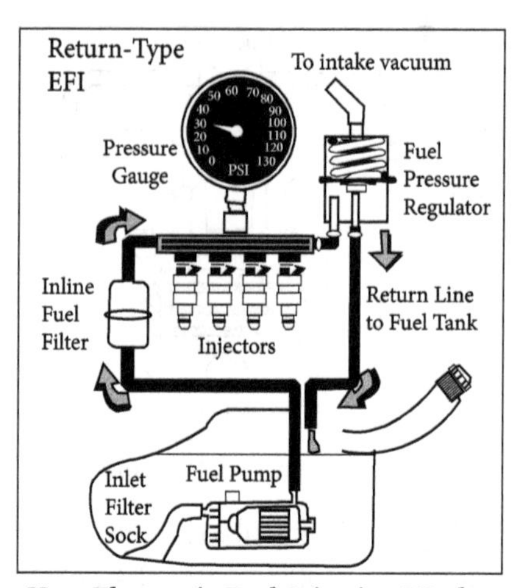

How Electronic Fuel Injection Works.

Electronic fuel injection (EFI) replaced carburetors back in the mid-1980s as the preferred method for supplying air and fuel to engines. The basic difference is that a carburetor uses intake vacuum and a pressure drop in the venturi (the narrow part of the carburetor throat) to siphon fuel from the carburetor fuel bowl into the engine whereas fuel injection uses pressure to spray fuel directly into the engine.

1.3.3 CRDI

The Fuel Injection in Compression Ignition Engines is in the form of Common Rail Direct Injection (CRDI) system shown in figure. This is a reliable system and can be adopted for most existing diesel engines after suitable changes.

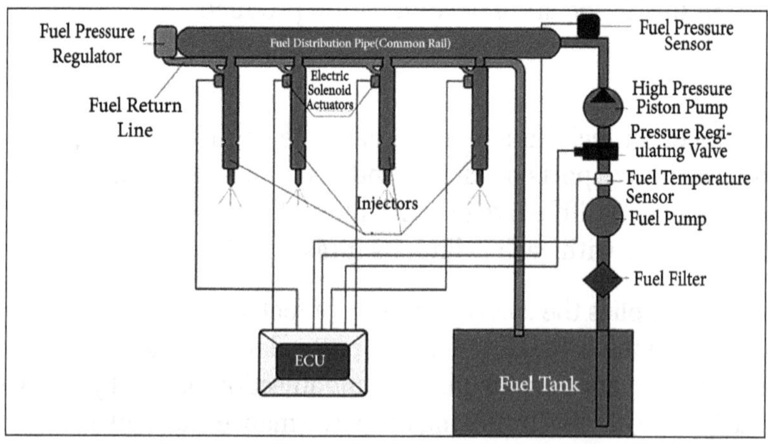

(a) Common Rail Direct Injection (CRDI) System.

The common rail is a large manifold and is continuously fed by fuel under pressure with the help of a pump driven by engine. The injectors are fed by pipes connected to rail. The injector is opened electrically. The injection timing is independent of engine cycle.

Diesel is a lower quality fuel and has particles larger and heavier than petrol. It is difficult to pulverize them. If pulverization is not proper the combustion of fuel leaves behind unburnt particles and more pollutants. This means lower fuel efficiency and less power. Common rail technology improves the pulverization of fuel. Here, a separate pump is used and fuel at high pressure is fed to individual fuel injector through a common pipe (common rail).

Fuel always remains at high pressure and therefore whenever the injector opens, high pressure fuel can be injected into combustion chamber quickly. As a result, along with improvement in pulverization, timing of fuel injection can be precisely controlled.

Working Principle

Combustion of fuel in an engine affects its overall performance. Combustion cannot occur in the absence of oxygen and that too is needed in a particular quantity. The source of oxygen being the fuel and air, form very important inputs for an engine. The combustibility of diesel is poorer than petrol therefore optimizing its mixing with air, injection into cylinder and burning is a complex process. Also the requirement of quantity of fuel to be mixed with air varies according to operating conditions.

In CRDI engines generating pressure and maintaining a real time check on amount of fuel injected have been separated. In this system, the common rail, which is a pipe, acts as a shared reservoir for electronically controlled injectors? The pressure of about 1500 bar is maintained in the common rail. With such high pressure built up in common rail the need for building up pressure in each injector is eliminated.

Connectors from common rail deliver diesel at high pressure to each injector. At the end of the injector, a solenoid Valve regulates the injection timing and amount of fuel to be injected on the basis of inputs from an electronic control module.

As the fuel is available at high pressure independent of operating conditions of engine it makes the engine better fuel efficient. The fuel is sprayed at high pressure, it ignites in the form of a controlled, yet violent explosion. The combustion is complete and that enhances the power output and clean emission.

The violent, explosion accompanying the combustion process, produces noise and vibration. To avoid this most CRDI engines employ 'Pilot Injection' or 'Pilot Burn'. A small amount of diesel is injected just before the main fuel injection. This starts the process of combustion before the injection of main fuel. This helps to make the explosion less violent. The rise in temperature and pressure is staggered and makes the operation of engine less noisy.

In modern diesel engines, ECM plays a very important role. Each injector has a solenoid valve that adjusts the injection timing as well as the amount of fuel to be injected. The electronic control module is speedy, reliable and can be programmed to take into account a large number of operating conditions. A mechanical device does not have all these qualities.

Advantages

1. CRDI engines are advantageous in many ways. Cars fitted with this new engine technology are believed to deliver 25% more power and torque than the normal direct injection engine.
2. It also offers superior pick up, lower levels of noise and vibration, higher mileage, lower emissions, lower fuel consumption and improved performance.
3. In India, diesel is cheaper than petrol and this fact adds to the credibility of the common rail direct injection system.

Disadvantages

1. This engine also has few disadvantages. The key disadvantage of the CRDI engine is that it is costly than the conventional engine.
2. The list also includes high degree of engine maintenance and costly spare parts. Also this technology can't be employed to ordinary engines.

1.3.4 GDI

Gasoline direct injection (GDI) can reduce charge-air temperature while allowing for higher compression ratios. This has the effect of reducing the potential for detonation yet increasing gasoline engine efficiency.

Instead of fuel and air mixing prior to entering the cylinder as with typical fuel injection, GDI uses a high-pressure injector nozzle to spray gasoline directly into the combustion chamber. Example of a GDI system is shown in figure. One advantage of GDI is that as the fuel vaporizes, it absorbs energy from the charge.

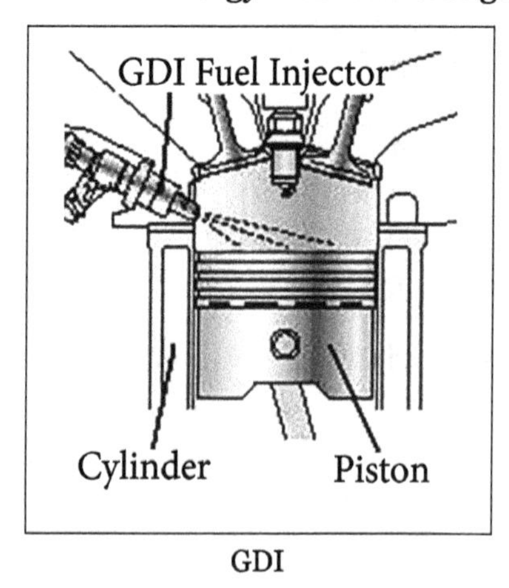

GDI

This "cooling effect" lowers the temperature of the air in the cylinder, thereby reducing its tendency to detonate.

Gasoline Direct Engines offer more advantages as compared to PFI engines, With regard to efficiency and specific power. To fully exploit this potential a particular attention should be paid to the in-cylinder formation process of air or fuel mixture. More demanding performance is required to the combustion system, since injectors should provide a fine fuel atomization in considerably short time, achieving a spray pattern to interact with in-cylinder air motion and piston top surface.

This is made possible through the common rail technology allowing an injection pressure one order of magnitude higher as compared with that of conventional PFI engines. Fuel economy can be obtained by regulating load, by mixture leaning, minimizing throttle usage at low loads where pumping losses are more significant and requiring charge stratification for a stable ignition and combustion.

Charge stratification can be pursued based mainly on the sole action of the fuel spray or on its interaction with a specially shaped surface on piston top or with the air bulk motion. Depending on the modality of stratification attainment, different combustion systems can be considered.

The injector design has in turn a key role being the final element of fuel metering required to the desired spray pattern, injected fuel mass per injection event, resistance to thermal stress and deposits. Injector housing and orientation with respect to the combustion chamber has to be carefully chosen, exploiting in this regard the indications of (CFD) computational fluid dynamics provided by 3D simulations.

1.3.5 HCCI

In HCCI engine, the homogeneous mixture of fuel and air is compressed and combustion begins when the appropriate conditions for auto-ignition are reached. This means that there is no well-defined combustion initiator such as spark plug as in a conventional SI engine. In HCCI gasoline fuelled engine, the density and temperature of the mixture are raised sufficiently by compression until the entire mixture reacts spontaneously without using electric discharge.

The characteristic of HCCI is that the ignition occurs at several places which makes the fuel/air mixture burn nearly simultaneously. There is no direct initiation of combustion. This makes the process challenging to control. However, with advances in microprocessors and a physical understanding of the ignition process, HCCI engine can be controlled.

To achieve dynamic operation in an HCCI engine, the control system must change the conditions that induce combustion. Thus, the engine must control the compression ratio, inducted gas temperature, inducted gas pressure, fuel-air ratio or quantity of retained or reinducted exhaust. To control the heat release rate to acceptable levels, the engine should be operated with high levels of dilution (i.e., with lean fuel/air ratio).

As the dilution levels are high, engine is operated un throttled, which reduces the pumping work and improves fuel economy. Fuel can be injected by using pott fuel injection (PFI) or direct injection (DI) technique. The large amounts of charge dilution also reduce the peak burned gas temperature, which in turn reduces the heat losses and increases the indicated thermal efficiency.To obtain optimum auto ignition timing, it is controlled by the pressure, temperature and com-position history of the charge during the compression process.

Thermal conditions and composition of the charge at intake valve closing must be correct to avoid a misfire. At lighter loads, ignition Is difficult to observe. At higher loads, the rate of pressure rise can become so large that engine noise increases and if left unchecked, engine damage may occur. Ultimately, the ability to control the combustion phasing at light loads and the bulk energy release rate at high loads will be key to successful HCCI engine development.

Compression Ratio

Raising the compression ratio would seem to be an obvious solution to triggering auto ignition of gasoline at near idle loads where fuel/air ratio is lean. However, it may cause severe knock problems and combustion generated noise at full load. Variable compression ratio would be a potential solution to this problem.

Fuel-Infection System

To obtain the most homogeneous mixture, it is desirable to have a long mixing timing between the fresh charge and the fuel. Thus, it would seem that early injection using a conventional port fuel injection system would be the most advantageous to obtain good homogeneous HCCI combustion. However, PFI injection offers no potential for additional combustion phasing control and limits the maximum usable compression ratio.

A switch to direct injection offers the potential for increased compression tabs and thus an extension of the HCCI light load limit. Direct injection also offers the potential for combustion phasing control by changing injection timing. By altering the injection timing from early in the intake stroke to late in the compression stroke, it is possible to obtain optimum combustion phasing over a range of intake air temperature, engine loads and speeds.

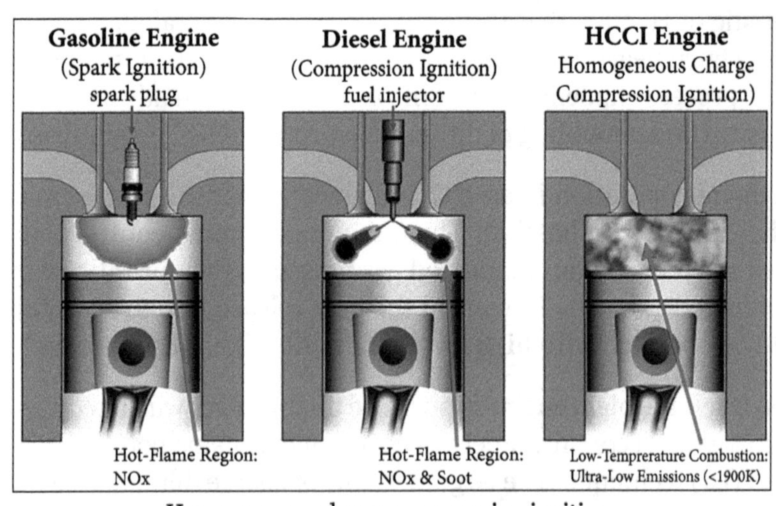

Homogeneous charge compression ignition.

Exhaust Gas Recirculation

The main challenge for HCCI operation is to obtain sufficient thermal energy to initiate auto ignition of the mixture late in the compression stroke at lower load. The most practical means to do this in a gasoline HCCI engine, where compression ratio is limited, is through the use of high levels of re-circulated exhaust gases. Above mid-load, the use of re-circulated exhaust gases is limited to control the energy release process.

Advantages

The following are the advantages of a homogeneous charge compression ignition engine:

1. HCCI provides up to 30% fuel savings while meeting current emissions standards.
2. Since HCCI engines are fuel-lean, they can operate at a diesel-like compression ratio, thus achieving higher efficiencies from conventional SI gasoline engines.
3. Homogeneous mixing of fuel and air leads to lower emissions. Since peak temperatures are significantly lower than those in typical SI engines, NOx levels are almost negligible. In addition, the premixed lean mixture does not produce soot.
4. HCCI engines can operate on gasoline, diesel fuel and most alternative fuels.
5. Compared to conventional SI engines, the absence of throttle losses improves HCCI efficiency.
6. In HCCI, the entire reactant mixture ignites almost simultaneously. Since very little pressure differences exist between the different regions of the gas, there is no shock wave propagation and hence no knocking.

1.3.6 Dual Fuel Engine

The way the gas is introduced depends on the engine design. It can be introduced upstream of the intake valve at the carb, as more industrial gas engines do today. It may be introduced and mixed with air at the beginning of the compression stroke before pilot diesel injection. This is called as fumigation.

It can also be directly injected under maximum pressure into the cylinder, like common rail direct fuel injection systems do on truck engines. Whenever ignition is triggered upon a small, pilot injection of diesel fuel set to the timing of the engine. The fumigated type of dual fuel engine arrangement is shown in below the figure:

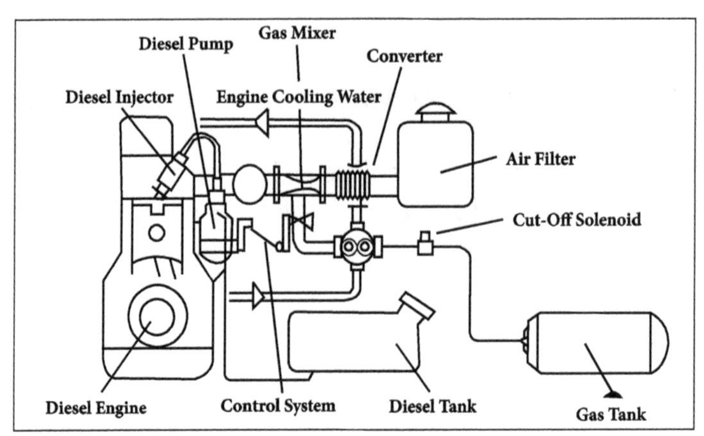

Dual fuel engine.

In oil fields, we have the flexibility of running an engine on cheap, widely available natural gas that would otherwise be flared, while having the torque curve approaching a diesel engine operation. Since its compression ignition, it can operate at high air fuel ratios and even be turbocharged suitably, since the diesel engine construction is a pretty solid one for thermal and mechanical loading.

The added advantage is of operating predominantly on gas which supports a lower combustion delay, because the fuel is gaseous. Another advantage is that during start-ups, when well head gas pressures are low to nil, diesel can be used to get the engine to run at normal speed.

Another advantage is on the emissions front, with low NOx and HC in the exhaust since its predominantly clean gaseous combustion. Because two fuels are used, the limitation maybe on the LHV of the least energetic gas. OEM's can also be particular about fuel gas composition and fuel quality. H_2S and CO_2 can lead to corrosive combustion by-products, therefore there's a limit on permissible compositions in these pollutants. It's always good to check with the engine application engineer on the predicted engine performance with a lower quality fuel and make sure a fuel conditioning system wouldn't be needed.

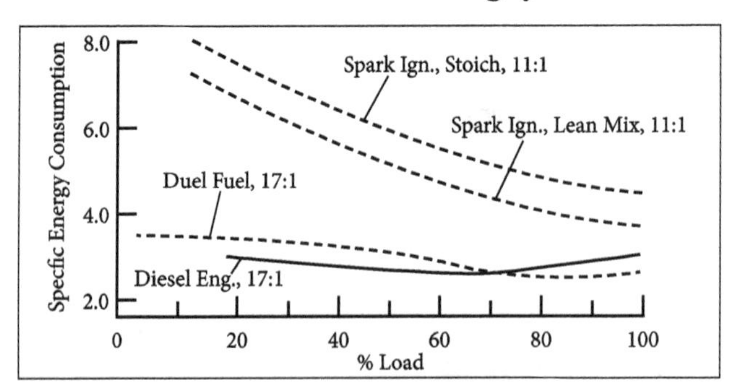

A comparison of variations of the specific energy consumption with load for spark ignition, diesel and dual-fuel engines operating on natural gas.

A figure from one study of dual fuel operational characteristic compared to other conventional modes of operation is shown above. The primary modes of control of performance in dual fuel engines are through varying gaseous-air fuel mixture and control of the size of pilot injection.

Advantage

The following are the advantages of a dual-fuel engine:

1. Cheap gases available from various sources can be used as a main fuel.
2. Gaseous fuels do not leave residue after combustion, therefore the exhaust is clean. The air pollution from the exhaust of the engine is very much reduced.
3. The clean combustion of gaseous fuels results in reduced wear of engine parts.
4. The maximum pressure in the cylinder is less than that for a diesel engine, thereby reducing the blow by past piston. This reduces the contamination and consumption of the lubricating oil.
5. The engine can run on varied proportions of gaseous and liquid fuels. With gaseous fuels the diesel oil requirement is hardly 5%. The engine can also be easily run with diesel oil alone.

1.3.7 Lean Burn Engine

Lean combustion in an engine is one of the most promising methods for reducing the exhaust emission and improving the fuel economy. The problems associated with lean burn are:

1. It is impossible to operate an engine with a mixture leaner than an air/fuel ratio of 19 due to deteriorated ignitability.
2. The fuel consumption tends to increase because of the lower combustion speed and the deterioration of the flame propagation in combustion with a lean mixture.
3. Increased variation of combustion from cycle to cycle causes fluctuations in torque, thus resulting in poor drivability.

Toyota lean burn engine with a pre-chamber has been developed taking care of the above problems. Figure shows the configuration of the lean burn combustion chamber. A fresh mixture flows into the pre-chamber through an orifice during the compression stroke, result in strong eddies of mixture within the pre-chamber. A spark plug located at the orifice ignites and produces a flame kernel in the mixture flow. This flame kernel flows into the pre-chamber with the mixture flow and causes rapid combustion in the pre-chamber.

As a result, an optimum jet flame spouts into the main combustion chamber. This jet flame increases the combustion speed of the lean mixture thus improving the combustion. The pre-chamber is named the Turbulence Generating Pot (TGP), since its function is to generate turbulence in the main combustion chamber. In addition to the TGP, the carburettor and the exhaust manifold are also modified for use in the lean burn engine.

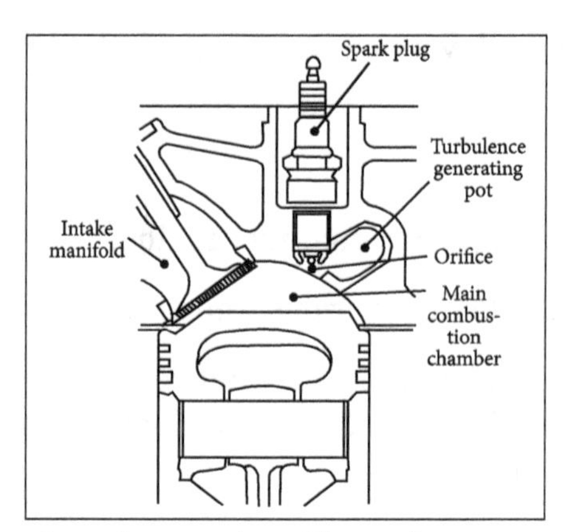

Learn burn engine with pre-champers.

The TGP improves the combustion of the lean mixture. The fuel consumption and torque fluctuations of the lean mixture combustion are reduced. Lean misfire limit is extended remarkably by locating the spark plug in the orifice of the TGP. Both the ignition lag and combustion noise are reduced due to the location of the spark plug.

1.3.8 Stratified Engine (Basic Principles)

The stratified charge engine is a type of internal combustion engine, used in automobiles, in which the fuel is injected into the cylinder just before ignition. This allows for maximum compression ratios without "knock," and leaner air and fuel ratio than in conventional internal combustion engines.

Conventionally, a four-stroke (petrol or gasoline) Otto cycle engine is fueled by drawing a mixture of air and fuel into the combustion chamber during the intake stroke. This produces a homogeneous mixture of air and fuel, which is ignited by a spark plug at a predetermined moment near the top of the compression stroke.

In a homogeneous charge system, the air and fuel ratio is kept very close to stoichiometric, meaning it contains the exact amount of air necessary for a complete combustion of the fuel. This gives stable combustion, but it places an upper limit on the engine's efficiency, any attempt to improve fuel economy by running a lean mixture with a homogeneous charge results in unstable combustion, this impacts on power and emissions, notably of nitrogen oxides or NOx.

1.4 Thermodynamic Analysis of Cycles: Significance of Fuel-Air and Actual Cycles of I.C. Engines

The actual cycles for IC engines differ from the fuel-air cycles and air-standard cycles in many respects. The actual cycle efficiency is much lower than the air-standard efficiency due to various Losses occurring in the actual engine operation.

The major losses are due to:

i. Variation of specific heats with temperature.

ii. Dissociation of the combustion products.

iii. Progressive combustion.

iv. Incomplete combustion of fuel.

v. Heat transfer into the walls of the combustion chamber.

vi. Blow down at the end of the exhaust proems.

vii. Gas exchange process.

An estimate of these losses can be made from previous experience and some simple tests on the engines and these estimates can be used in evaluating the performance of an engine.

Actual and Fuel-Air Cycles of IC Engines

In the diesel cycle the losses are less compared to Otto cycle. The main loss is due to incomplete combustion and is the cause of difference between fuel-air cycle and actual cycle of a diesel engine. In a fuel-air cycle the combustion is completed at the end of the constant pressure burning whereas in actual practice, after burning continues up to half of the expansion stroke. The ratio between the actual efficiency and the fuel-air cycle efficiency is about 0.85 in the diesel engines.

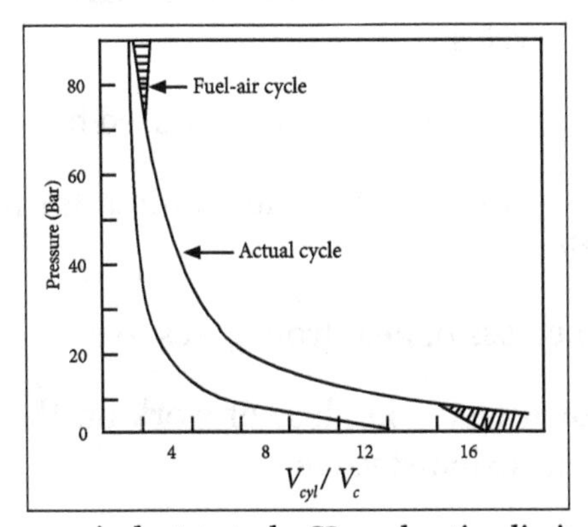

Actual fuel cycle vs equivalent 2 stroke SI combustion limited pressure cycle.

In fuel-air cycles, when allowance is made for the presence of fuel and combustion products, there is reduction in cycle efficiency. In actual cycles, allowances are also made for losses due to heat transfer and finite combustion time. This reduces the cycle efficiency further. For complete analysis of actual cycles, computer models are being developed nowadays. These models help us in understanding the various processes that are taking place in an engine.

1.4.1 Comparison with Air Standard Cycles

The actual cycles for internal combustion engines differ from air-standard cycles in many respects. These differences are mainly due to,

i. The working substance being a maximum of air and fuel vapour or finely atomized liquid fuel in air combined with the products of combustion left from the previous cycle.

ii. The change in chemical composition of the working substance.

iii. The different specific heats with temperature.

iv. The change in the composition temperature and actual amount of fresh charge because of the residual gases.

v. The progressive combustion rather than the instantaneous combustion.

vi. The heat transfer to and from the working medium.

vii. The substantial exhaust blow down loss, i.e., loss of work on the expansion stroke due to early opening of the exhaust valve.

viii. Gas leakage, fluid friction etc., in actual engines.

Points (i) to (iv), being related to fuel-air cycles. Remaining points viz. (v) to (viii) are in fact responsible for the difference between fuel-air cycles and actual cycles. Most of the factors listed above are to decrease the thermal efficiency and power output of the actual engines. On the other hand, the analysis of the cycles while taking these factors into account clearly indicates that the estimated thermal efficiencies are not very different than those of the actual cycles.

Out of all the above factors major influence is exercised by,

i. Time loss factor i.e. loss due to time required for mixing of fuel and air and also for combustion.

ii. Heat loss factor i.e. loss of heat from gases to cylinder walls.

iii. Exhaust blow down factor i.e. loss of work on the expansion stroke due to early opening of the exhaust valve.

1.5 Analysis of Fuel-Air and Actual Cycles

i. In air standard cycles analysis highly simplified approximations are made. The air standard theory gives an estimate of engine performance which is much greater than the actual performance. This large variation is partly due to the non-instantaneous burning and valve operation, incomplete combustion etc. the major reason being over complication in using the properties of the working fluid for cycle analysis.

ii. In air cycle approximation it is assumed that air is a perfect gas having constant specific heats. In actual engine the working fluid is not air but a mixture of air, fuel and exhaust gases. Furthermore, the specific heats of the working fluid are not constant but increase with rise in temperature and at high temperature the combustion products are subjected to dissociation.

iii. The theoretical cycle based on the actual properties of the cylinder the Fuel-air cycle approximation, it provides a rough idea for comparison with the actual performance.

Importance of Fuel-Air Cycle

Whereas the air standard cycle exhibits the general effect of compression ratio on efficiency of the engine, the fuel-air cycle may be calculated for various fuel-air ratios, inlet temperatures and pressures (It is worth noting that fuel-air ratio and compression ratio are much more important parameters in comparison to inlet conditions).

With the help of fuel-air cycle analysis a very good estimate of power to be expected from the actual engine can be made. Furthermore, it is possible to approximate very closely peak pressures and exhaust temperatures on which design and engine structure depend.

Variable Specific Heats

All gases, except mono-atomic gases, show an increase in specific heat with temperature. The increase in specific heat does not follow any particular law. However, for the temperature range generally encountered for gases in heat engines (300 K to 2000 K) the specific heat curve is nearly a straight line which may be approximately expressed in the form,

$$\left.\begin{aligned} C_p &= a_1 + k_1 T \\ C_v &= b_1 + k_1 T \end{aligned}\right\} \quad ...(1)$$

Where, a1, b1 and k1 are constants. Now,

$$R = C_p - C_v = a_1 - b_1 \quad ...(2)$$

Where, R -Characteristic gas constant.

Above 1500 K the specific heat increases much more rapidly and may be expressed in the form,

$$Cp = a_1 + k_1 T + k_2 T^2 \quad ...(3)$$

$$Cv = b_1 + k_1 T + k_2 T^2 \quad ...(4)$$

In Equation 4 if the term k_2T^2 is neglected it becomes same as equation 1. Many expressions are available even up to sixth order of T (i.e. T^6) for the calculation of C_p and C_v. The physical explanation for increase in specific heat is that as the temperature is raised, larger fractions of the heat would be required to produce motion of the atoms within the molecules. Since temperature is the result of motion of the molecules, as a whole, the energy which goes into moving the atoms does not contribute to proportional temperature rise. Hence, more heat is required to raise the temperature of unit mass through one degree at higher levels. This heat by definition is the specific heat. The values for Cr and C for air are usually taken as,

$$C^p = 1.005 \text{ kJ / kg K at 300K}$$

$$C^p = 1.345 \text{ kJ / kg K at 2000K}$$

$$C^v = 0.717 \text{ kJ / kg K at 300K}$$

$$C^v = 1.057 \text{ kJ / kg K at 2000K}$$

Since, the difference between the value of γ decreases with increase in temperature. Thus, if the variation of specific heats is taken into account during the compression stroke, the final temperature and pressure would be lower if constant values of specific heat are used. This point is illustrated in Figure:

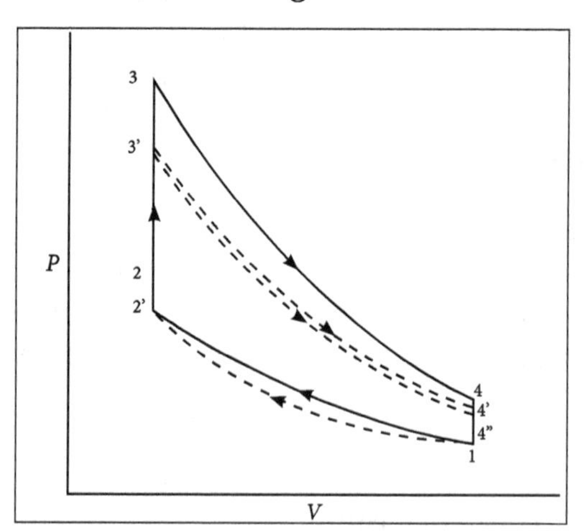

Loss of power due to variation of specific heat.

Cycle 1-2-3-4: With constant specific heat.

Cycle 1-2'-3'-4': With variable specific heat.

Cycle 1-2-3'-4": With constant specific heat from point 3'.

With variable specific heats, the temperature at the end of compression will be 2', instead of 2. The magnitude of drop in temperature is proportional to the drop in the value of ratio of specific heats. For the process $1 \rightarrow 2$, with constant specific heats,

$$T_2 = T_1\left(\frac{V_1}{V_2}\right)^{(\gamma-1)} \qquad ...(5)$$

With variable specific,

$$T_{2'} \quad T_1\left(\frac{}{_{2'}}\right)^{(k\ 1)} \qquad ...(6)$$

For given values of T_1, p_1 and r, the magnitude of T_2, depends on k. Constant volume combustion, from point 2' will give e temperature T_a, instead of T_0. This is due to the fact that the rise in the value of C because of variable specific heat, which reduces the temperature as already, explained. The process, $2' \rightarrow 3'$ is heat addition with the variation in specific heat.

From 3', if expansion takes place at constant specific heats, this would result in the process $3' \rightarrow 4''$ whereas actual expansion due to variable specific heat will result in $3' \rightarrow 4'$ and 4' is higher than 4". The magnitude M the difference between 4' and 4" is proportional to the reduction in the value of γ. Consider the process $3' \rightarrow 4''$

$$T_{4''} = T_{3'}\left(\frac{V_3}{V_4}\right)^{(k-1)} \qquad ...(7)$$

for the process $3' \rightarrow 4'$

$$T_{4'} = T_{3'}\left(\frac{V_3}{V_4}\right)^{(\gamma-1)} \qquad ...(8)$$

Reduction in the value of k due to variable specific heat results in increase of temperature from $T_{4''}$ to $T_{4'}$.

1.6 Effect of Air-Fuel Ratio

Engine Air-Fuel Ratio

An automobile SI engine, as indicated above, works with the air-fuel mixture ranging from 8:1 to 18.5:1. But the ideal ratio would be one that provides both the maximum power and the best economy, while producing the least emissions. But such a ratio does not exist because the fuel requirements of an engine vary widely depending upon temperature, load and speed conditions.

The best fuel economy is obtained with a 15:1 to 16:1 ratio, while maximum power output is achieved with a 12.5:1 to 13.5:1 ratio. A rich mixture in the order of 11:1 is required for idle heavy load and high-speed conditions. A lean mixture is required for normal cruising and light load conditions. Figure represents the characteristic curves showing the effect of mixture ratio on efficiency and fuel consumption.

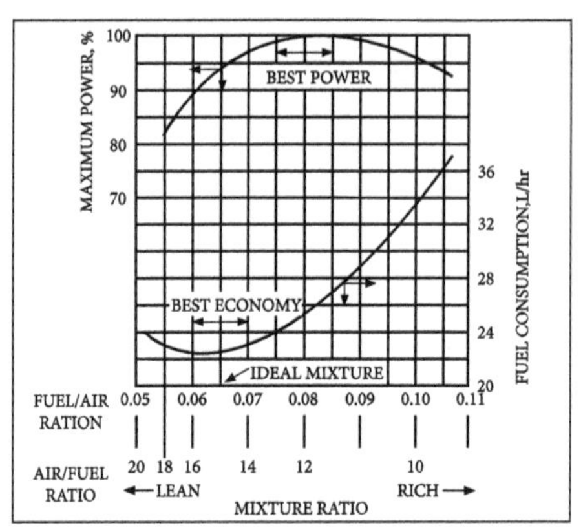

Effect of air-fuel ratio on efficiency and fuel consumption.

Practically for complete combustion, mixing of the fuel in excess air (to a limited extent above that of the ideal condition) is needed. Lean mixtures are used to obtain best economy through minimum fuel consumption whereas rich mixtures used to suppress combustion knock and to obtain maximum power from the engine.

However, improper distribution of mixture to each cylinder and imperfect/incomplete vaporization of fuel in air necessitates the use of rich mixture to obtain maximum power output. A rich mixture is also required to overcome the effect of dilution of incoming mixture due to entrapped exhaust gases in the cylinder and of air leakage because of the high vacuum in the manifold, under idling or no-load condition.

Maximum power is desired at full load while best economy is expected at part throttle conditions. Thus required air fuel ratios result from maximum economy to maximum power. The carburettor must be able to vary the air-fuel ratio quickly to provide the best possible mixture for the engine's requirements at a given moment. The best air-fuel ratio for one engine may not be the best ratio for another, even when the two engines are of the same size and design. To accurately determine the best mixture, the engine should be run on a dynamometer to measure speed, load and power requirements for all types of driving conditions. With a slightly rich mixture, the combustion flame travels faster and conversely with a slightly weak mixture, the flame travel becomes slower. If a very rich mixture is used then some "neat" petrol enters cylinder, washes away lubricant from cylinder walls and gets past piston to contaminate engine oil. A very sooty deposit occurs in the combustion chamber.

On the other hand, if an engine runs on an excessively weak mixture, then overheating particularly of parts such as valves, pistons and spark plugs occurs. This causes detonation and pre-ignition together or separately. The approximate proportions of air to petrol (by weight) suitable for the different operating conditions are indicated below:

Starting	9	1
Idling	12	1
Acceleration	12	1
Economy	16	1
Full power	12	1

It makes no difference if an engine is carburetted or fuel injected, the engine still needs the same air-fuel mixture ratios.

1.6.1 Exhaust Gas Dilution

There are a number of techniques used to characterize and measure particulate emissions. These include light absorption, filter discoloration, and measurement of the total mass of particulates trapped on a filter paper. The exhaust particle size distribution can be measured using aerosol instruments such as the scanning mobility particle size (Wang and Plague, 1990). The absorption-type smoke meter uses the principle of light absorption by particles.

A pump is used to draw undiluted exhaust gas into a measuring chamber that has a light source at one end and a photodiode at the opposite end. The attenuation of the beam of light by the exhaust is proportional to the particle concentration. The filter-type smoke meter draws a metered amount of exhaust gases through a filter paper.

The blackening of the filter paper is compared against a "Bacharach grey scale." Standards, such as SAE J 1280, that employ direct mass measurement also specify the use of a dilution tunnel in order to simulate the exhaust conditions near a vehicle. The particulates leaving the exhaust pipe are at a relatively high-temperature and concentration in the outlet exhaust flow. These gases cool during the mixing process with the atmosphere, and the associated condensation and agglomeration processes will change the structure and density of the particulates in the exhaust gases.

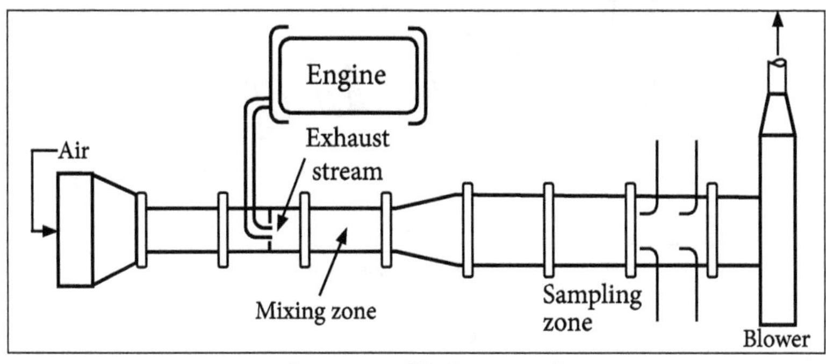

Exhaust gas dilution tunnel.

Dilution tunnels are used to standardize this near-field mixing process. A dilution tunnel is shown in figure above. The tunnel is about 0.3 mm in diameter. By flowing dilution air at a constant speed, typically 10 m/s, through a converging diverging nozzle, the Venturi effect can be used to remove exhaust gas from the exhaust pipe. Mini dilution tunnels with a 25-mm diameter have also been developed. Downstream of the nozzle, the exhaust is well mixed with the dilution air.

The diluted exhaust gas is sampled and drawn through Teflon-coated glass fiber paper filters. The total particulate mass is trapped by the filter found by the increase in weight of the sample filter. In order to compute the dilution ratio, which is defined as the ratio of dilute mixture flow rate to exhaust gas flow rate, the carbon dioxide concentration measured in both the engine exhaust and the diluted sample. The dilution ratio is typically about 10:1.

1.7 Time and Heat Loss Factor

Time Loss Factor

In air-standard cycles the heat addition is assumed to be an instantaneous process whereas in an actual cycle it is over a definite period of time. The time required for the combustion is such that under all circumstances some change in volume takes place while it is in progress. The crankshaft will usually turn about 30° to 40° between the initiation of the spark and the end of combustion. There will be a time loss during this period and is time loss factor.

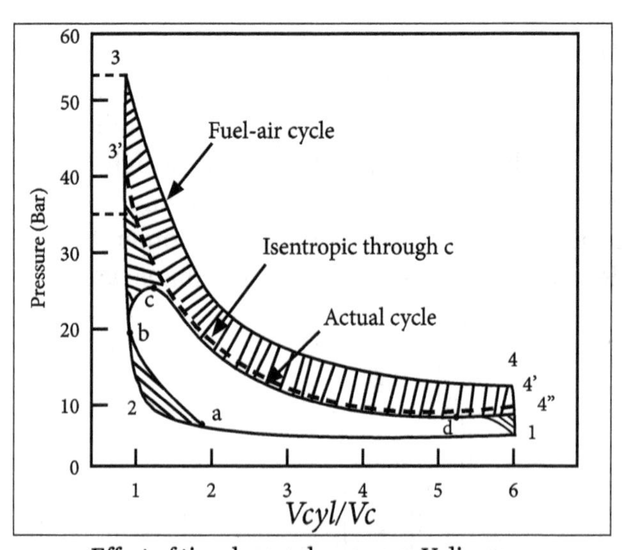

Effect of time losses shown on p-V diagram.

The consequence of the finite time of combustion is that the peak pressure will not occur when the volume is minimum i.e., when the piston is at TDC, but will occur sometime after TDC. The pressure, therefore, rises in the first part of the working stroke from b to c , as shown in figure the point three represents the state of gases had the combustion been instantaneous and an additional amount of work equal to area shown hatched would have been done. This loss of work reduces the efficiency and is called time loss due to progressive combustion or merely time losses.

The time taken for the burning depends upon the flame velocity which in turn depends upon the type of fuel and the fuel-air ratio and also on the shape and size of the combustion chamber. Further, the distance from the point of ignition to the opposite side of the combustion space also plays an important role.

Cycle performance for various ignition timing for r=6

Cycle	Ignition advance	Max. Cycle pressure bar	Meo bar	Efficiency %	Actual η/Fuel cycle η
Fuel-air cycle	0°	44	10.20	32.2	1.00
Actual cycle	0°	23	7.50	24.1	0.75
Actual cycle	17°	34	8.35	26.3	0.81
Actual cycle	35°	41	7.61	23.9	0.74

Heat Loss Factor

During the combustion process and the subsequent expansion stroke the heat flows from the cylinder gases through the cylinder walls and cylinder head into the water jacket or cooling fins. Some heat enters the piston head and flows through the piston rings into the cylinder well or is carried away by the engine lubricating oil which splashes on the underside of the piston. The heat loss along with other losses is shown on the p-V diagram in figure.

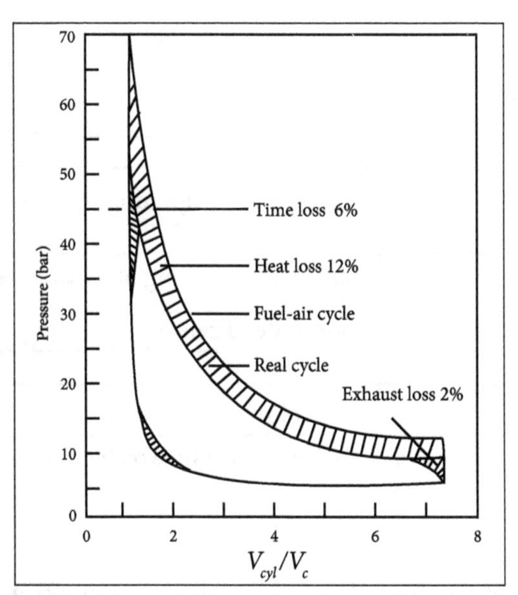

Time loss, heat loss and exhaust loss in petrol in petrol engine.

Heat loss during combustion will naturally have the maximum effect on the cycle efficiency while heat loss just before the end of the expansion stroke can have very little effect because of its contribution to the useful work is very little. The heat lost during the combustion does not represent a complete loss because, even under ideal conditions assumed for air-standard cycle, only a part of this heat could be converted into work and the rest would be rejected during the exhaust stroke. About 15 per cent of the total heat is lost during combustion and expansion. However, much is lost so late in the cycle to have contributed to useful work.

If all the heat loss is recovered only about 20% of it may appear as useful work. Figure shows percentage of time loss, heat loss and exhaust loss in a Cooperative Fuel Research (CFR) engine. Losses are given as percentage of fuel-air cycle work. The effect of loss of heat during combustion is to reduce the maximum temperature and therefore, the specific heats are lower. It may be noted from the figure that of the various losses, heat loss factor contributes around 12%.

1.7.1 Exhaust Blow down

The cylinder pressure at the end of exhaust stroke is about 7 bar depending on the compression ratio employed. If the exhaust valve is opened at the bottom dead centre, the piston tends to do work against high cylinder pressures during the early part of the exhaust stroke.

If the exhaust valve is opened too early, a part of the expansion stroke is lost. The best compromise is to open the exhaust valve 90° to 70° before BDC thereby reducing the cylinder pressure to halfway before the exhaust stroke begins.

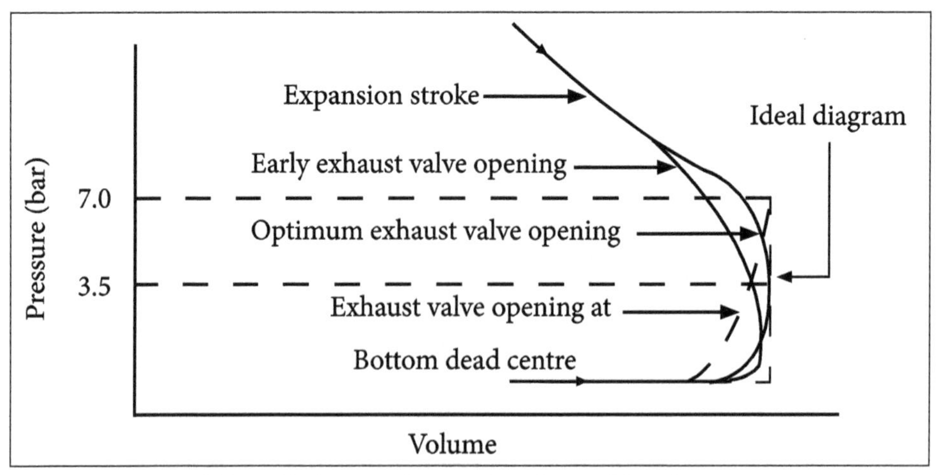

Effect of exhaust valve opening time on blow down.

Blowby loss

It is due to the leakage of gas flow through crevices or gaps present in between piston, piston rings and cylinder walls. The gas usually leaks through the crankcase which is often called as crankcase blowby.

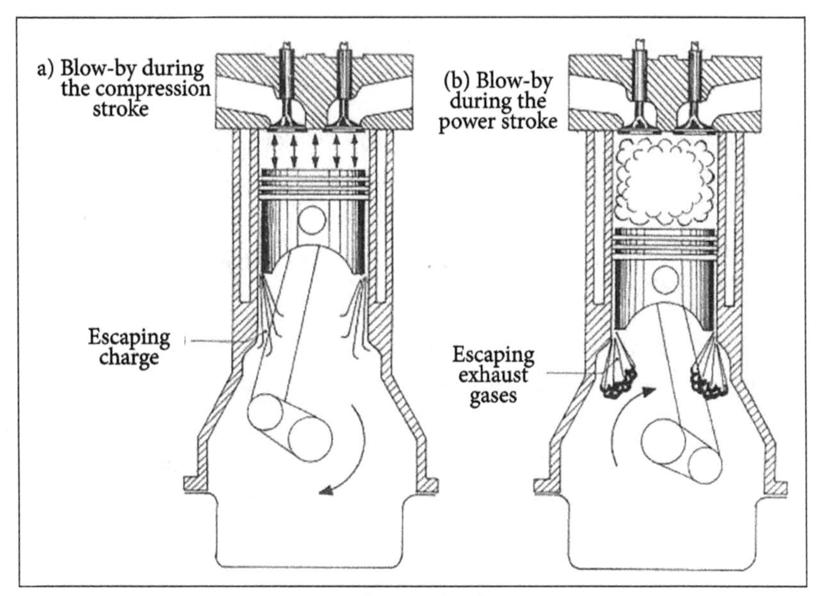

Crankcase Blowby.

1.7.2 Loss Due to Gas Exchange Processes

The difference of work done in expelling the exhaust gases and the work done by the fresh charge during the suction stroke is called the pumping work. In other words loss due to the gas exchange process (pumping loss) is due to pumping gas from lower inlet pressure pe to higher exhaust pressure p_i.

The pumping loss increases at part throttle because throttling reduces the suction pressure. Pumping loss also increases with speed. The gas exchange processes affect the volumetric efficiency of the engine. The performance of the engine, to a great deal, depends on the volumetric efficiency. Hence, it is worthwhile to discuss this parameter in greater detail here.

1.7.3 Volumetric Efficiency

Volumetric efficiency is the indication of the breathing ability of the engine and is defined as the ratio of the volume of air actually inducted at ambient condition to swept volume. However, it may also be defined as the ratio of the actual mass of air drawn into the engine during a given period of time to the theoretical mass which should have been drawn in during that same period of time, based upon the total piston displacement of the engine and the temperature and pressure of the surrounding atmosphere.

The volumetric efficiency of a compressor is the ratio of free air delivered to the displacement of the compressor. It is also the ratio of effective swept volume to the swept volume.

$$\text{i.e., Volumetric efficiency} = \frac{\text{Effective swept volume}}{\text{Swept volume}} = \frac{V_1 - V_4}{V_1 - V_3}$$

Because of presence of clearance volume, volumetric efficiency is always less than unity, as a percentage, it usually varies from 60% to 85%.

The ratio, $\frac{\text{Clearance volume}}{\text{Swept volume}} = \frac{V_3}{V_1 - V_3} = \frac{V_c}{V_s} = k$ is the clearance ratio.

As a percentage, this ratio will have a value between 4% and 10%. The greater the pressure ratio through a reciprocating compressor, then the greater will be the effect of the clearance volume since the clearance air will now expand through a greater volume before intake conditions are reached. The cylinder size and stroke being fixed, however will mean that (V1 - V4), the effective swept volume, will reduce as the pressure ratio increases and thus the volumetric efficiency reduces.

Volume efficiency, $\varsigma_{vol} = \frac{V_1 - V_4}{V_1 - V_3}$

$$= \frac{(V_1 - V_3) - (V_3 - V_4)}{(V_1 - V_3)} = 1 + \frac{V_1}{V_1 - V_3} - \frac{V_4}{V_1 - V_3}$$

$$= 1 + \frac{V_3}{V_1 - V_3} - \frac{V_4}{V_1 - V_3} \cdot \frac{V_3}{V_3} = 1 + \frac{V_3}{V_1 - V_3} - \frac{V_3}{V_1 - V_3} \cdot \frac{V_4}{V_3}$$

$$= 1 + k - k \cdot \frac{V_4}{V_3} \quad \left| \begin{array}{l} p_3 V_3^n = p_4 V_4^n \\ \frac{V_4}{V_3} = \left(\frac{p_3}{p_4}\right)^{1/n} \end{array} \right.$$

$$= 1 + k - k\left(\frac{p_3}{p_4}\right)^{1/n}$$

$$\text{Or } \eta_{vol} = 1 + k - k\left(\frac{p_2}{p_1}\right)^{1/n} \quad (\because p_3 = p_2, p_4 = p_1)$$

$$\text{Or } \eta_{vol} = 1 + k - k\left(\frac{V_1}{V_2}\right)$$

The above equations are valid if the index of expansion and compression is same. However it may be noted that the clearance volumetric efficiency is dependent on only the index of expansion of the clearance volume from V3 to V4. Thus, if the index of compression = nc and index of expansion = ne, the volumetric efficiency is given by,

$$\eta_{vol} = 1 + k - k\left(\frac{p_3}{p_4}\right)^{1/n_c}$$

$$= 1 + k - k\left(\frac{p_2}{p_1}\right)^{1/n_c}$$

$$= 1 + k - k\left(\frac{V_4}{V_3}\right)$$

In this case volumetric efficiency $= 1 + k - k\left(\frac{V_1}{V_2}\right)$

1.7.4 Loss Due to Rubbing Friction

These losses are due to friction between the piston and the cylinder walls, friction in various bearings and also the energy spent in operating the auxiliary equipment such as cooling water pump, ignition system, fan, etc. The piston ring friction increases rapidly with engine speed. It also increases to a small extent with increase in mean effective pressure.

The bearing friction and the auxiliary friction also increase with engine speed. The efficiency of an engine is maximum at full load and decreases at part loads. It is because the percentage of direct heat Ices, pumping loss and rubbing friction loss increase at part loads.

The approximate losses for a gasoline engine of high compression ratio, say 8:1 using a chemically correct mixture is given in Table, as percentage of fuel energy input.

Typical losses in a gasoline engine for r =8.

S. No.	Item	At load	
		Full load	Half load
(a)	Air-standard cycle efficiency ($\eta_{air-std}$).	56.5	56.5
1	Losses due to variation of specific heat and chemical equilibrium, %	13.0	13.0
2	Loss due to progressive combustion, %	4.0	4.0
3	Loss due to incomplete combustion, %	3.0	3.0
4	Direct heat loss, %	4.0	5.0
5	Exhaust blow down loss, %	0.5	0.5
6	Pumping loss, %	0.5	1.5
7	Rubbing friction loss, %	3.0	6.0
(b)	Fuel-air cycle efficiency = $\eta_{air-std} - (1)$	43.5	43.5

(c)	Gross indicated thermal efficiency (η_{th}) = Fuel-air cycle efficiency (η_{ith}) $-(2+3+4+5)$	32.0	31.0
(d)	Actual brake thermal efficiency $=(\eta_{ith})-(6+7)$	28.5	23.5

1.8 Fuels of SI and CI Engine

General Aspects

The crude oil and petroleum products, sometimes during the 21st century will become very scarce and costly to fluid and produces. At the same time, there will likely be an increase in the number of automobiles and other I.C engines. Although fuel economy of engines is greatly improved from the past and will probably continue to be improving only dictate that there will be a great demand for fuel in the coming decades.

Gasoline will become scarce and costly. Alternate fuel technology, will become more common in the coming decades. Because of High Cost of petroleum products, some third world countries have for many years been using manufacturing alcohol as their main vehicle fuel. Another reason motivating the development of alternative fuels for the I.C engine is concern over the emission problems of gasoline engines.

Combined with an air polluting systems, the large number of automobile is a major contributor to the air quality problem of the world. Vast improvements have been made in reducing emissions given off by all automobile engines.

Some alternate fuels which can replace conventional fuels in I.C engines are,

1. Alcohol (Methyl & Ethyl)
2. Hydrogen
3. Natural gas
4. LPG and LNG
5. Bio gas

Still another reason for alternate fuel development in India and other industrialized countries is the fact that a large percentage of crude oil must be supported from other countries which control the large oil fields.

Bio Diesel Production

Bio diesel is simply a liquid fuel derived from vegetable oils and fats, which has similar combustion properties to regular petroleum diesel fuel. Bio diesel can be produced from straight vegetable oil, animal oil or fats, and waste cooking oil. Bio diesel is biodegradable, nontoxic and has significantly fewer emissions than petroleum-based diesel when burned.

Bio diesel is an alternative fuel similar to conventional or "fossil and petroleum" diesel. The process used to convert these oils to bio diesel is called trans-esterification. This process is described in more detail below. The largest possible source of suitable oil comes from oil crops such as soybean, rapeseed, corn and sunflower.

At present, oil straight from the agricultural industry represents the greatest potential source, but it is not being used for commercial production of bio diesel simply because the raw oil is too expensive. After the cost of converting it to bio diesel has been added, the price is too high to compete with petroleum diesel.

Waste vegetable oil can often be obtained for free or already treated for a small price. One disadvantage of using waste oil is it must be treated to remove impurities like free fatty acids (FFA) before conversion to bio diesel. In conclusion, bio diesel produced from waste vegetable or animal's oil and fats can compete with the prices of petroleum diesel without national subsidies.

Making Bio Diesel: Trans-Esterification

Trans-esterification of natural glycerides with methanol to methyl-esters is a technically important reaction that has been used extensively in the soap and detergent manufacturing industry worldwide for many years. Almost all bio-diesel is produced in a similar chemical process using base catalyzed trans-esterification as it is the most economical process, requiring only low temperatures and pressures while producing a 98% conversion yield.

The trans-esterification process is the reaction of a triglyceride with an alcohol to form esters and glycerol. A triglyceride has a glycerin molecule as its base with three long chain fatty acids attached. The characteristics of the fat are determined by the nature of the fatty acids attached to the glycerin. The nature of the fatty acids can, in turn, affect the characteristics of the bio diesel.

During the esterification process, the triglyceride is reacted with alcohol in the presence of a catalyst usually a strong alkaline like sodium hydroxide. The alcohol reacts with the fatty acids to form the mono-alkyl ester or bio diesel and crude glycerol. In most production, methanol or ethanol is the alcohol used and is base catalyzed by either potassium or sodium hydroxide. Potassium hydroxide has been found more suitable for the ethyl ester bio diesel production, but either base can be used for methyl ester production.

The figure below shows the chemical process for methyl ester bio diesel. The reaction between the fat or oil and the alcohol is a reversible reaction, so the alcohol must be added in excess to drive the reaction towards the right and ensure complete conversion.

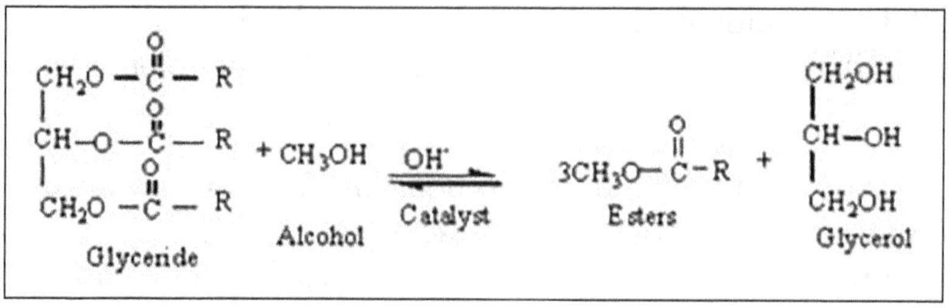

$$\underset{\text{Glyceride}}{\begin{matrix} CH_2O-\overset{O}{\overset{\|}{C}}-R \\ | \\ CH-O-\overset{O}{\overset{\|}{C}}-R \\ | \\ CH_2O-\overset{O}{\overset{\|}{C}}-R \end{matrix}} + \underset{\text{Alcohol}}{CH_3OH} \underset{\text{Catalyst}}{\overset{OH^-}{\rightleftharpoons}} \underset{\text{Esters}}{3CH_3O-\overset{O}{\overset{\|}{C}}-R} + \underset{\text{Glycerol}}{\begin{matrix} CH_2OH \\ | \\ CH-OH \\ | \\ CH_2OH \end{matrix}}$$

The products of the reaction are the bio diesel itself and glycerol.

A successful trans-esterification reaction is signified by the separation of the methyl ester (bio diesel) and glycerol layers after the reaction time. The heavier co-product, glycerol, settles out and may be sold as is or purified for use in other industries, example pharmaceutical, cosmetics and detergents.

After the trans-esterification reaction and the separation of the crude heavy glycerin phase, the producer is left with a crude light bio diesel phase. This crude bio diesel requires some purification prior to use.

Bio diesel has a viscosity similar to petroleum diesel and can be used as an additive in formulations of diesel to increase the lubricity. Bio diesel can be used in pure form blended with petroleum diesel at any concentration in most modern diesel engines.

Bio diesel will degrade natural rubber gaskets and hoses in vehicles although these tend to wear out naturally and most likely will have already been replaced with Viton type seals and hoses which are nonreactive to bio diesel. Bio diesel's higher lubricity index compared to petroleum diesel is an advantage and can contribute to longer fuel injector life.

Bio diesel is a better solvent than petroleum diesel and has been known to break down deposits of residue in the fuel lines of vehicles that have previously been run on petroleum diesel. Fuel filters may become clogged with particulates if a quick transition to pure bio diesel is made, as bio diesel "cleans" the engine in the process. It is, therefore, recommended to change the fuel filter within 600-800 miles after first switching to a bio diesel blend.

Bio diesel's commercial fuel quality is measured by the ASTM standard designated D 6751.The standards ensure that bio diesel is pure and the following important factors in the fuel production process are satisfied:

- Complete reaction.
- Removal of glycerin.
- Removal of catalyst.
- Removal of alcohol.
- Absence of free fatty acids.

Low Sulfur Content

Bio diesel is, at present, the most attractive market alternative among the non-food applications of vegetable oils for transportation fuels. The different stages in the production of plant/seed oil methyl ester generate by-products which offer further outlets. Oil cake, the protein rich fraction obtained after the oil has been extracted from the seed, is used for animal feed. Glycerol, the other important by-product, has numerous applications in the oil and chemical industries such as the cosmetic, pharmaceutical, food and painting industries.

1.8.1 Fuel Additives and Properties

The Importance of Fuel Additives

Many professional auto repair technicians often classify fuel additives as “mechanic in a can” solutions to various fuel delivery problems, but fuel additives can prevent or remedy many fuel delivery system problems. The topic of fuel additives is often clouded by a lack of definition between the different categories of fuel additives and the conditions they’re designed to address.

To avoid further confusion on the topic of fuel additives, we’ll discuss the need for various categories of fuel additives from a historical perspective. Most of these fuel additives will fall into seven basic categories:

1. Gasoline Stabilizers

Loss of a gasoline’s volatility isn’t a factor in normal driving, but it becomes a major problem when starting seasonal-use vehicles like boats, motor homes, lawn care equipment and electric generators. Using a gasoline stabilizer basically reduces hard-starting caused by the tendency of the more gaseous components of gasoline to evaporate into the atmosphere. Many new formulas now include corrosion inhibitors that prevent fuel system corrosion from ethanol-based gasoline.

2. Fuel-Line Antifreeze

From the earliest days of the automobile, water condensation normally found in gasoline storage tanks and in automobile fuel tanks often caused the fuel line to freeze during cold weather. The historic remedy is to add various alcohol-based fuel line antifreezes that will mix with the water to prevent fuel line freeze-up. On another level, water dispersant fuel line anti-freezes will actually absorb and transport water from the gasoline through the engine, where it is vaporized into the exhaust stream. Although modern ethanol fuels perform the same function, many modern fuel additives nevertheless include some type of water dispersant in their various formulas.

3. Octane Boosters

Ethyl lead was used in gasoline for many years to increase the gasoline's octane rating and lubricate engine valve seats. After the use of ethyl lead was discontinued during the 1970s, octane booster additives became popular for increasing the octane rating of gasoline used in the high-compression engines of the day. In most cases, a valve seat lubricant is included in an octane booster to reduce wear on antique and collector car engines equipped with cast-iron valve seats.

4. Fuel Injector Cleaners

The symptoms of clogged fuel injectors are hard-starting, poor cold-engine performance and sluggish acceleration. These symptoms are caused by carbon deposits clogging the fuel injector nozzles. Most lower-priced fuel injector cleaners are designed as preventive maintenance additives while the higher-priced cleaners are designed to also remove heavy carbon deposits from cylinder heads, valves and pistons. Any of these additives generally perform well when used as directed.

5. Upper Cylinder Lubricants

Thanks to the use of high-viscosity engine oils, varnish deposits and low-speed driving, early L-head or "flat head" gasoline engines often had problems with engine valves sticking in their valve guides. Light-viscosity, high-detergent upper cylinder oils were designed to be added to the fuel tank to help clean and lubricate sticking valves. Nowadays, sticking valves are rare and usually caused by insufficient oil clearance in the valve guide assembly. Nevertheless, many modern gasoline additives contain some type of upper cylinder lubricant to lubricate valve guides, seats and piston rings.

6. Diesel Exhaust Fluid

While diesel exhaust fluid (DEF) isn't a fuel additive, it's important to understand how it's used. DEF is metered from a separate tank on the vehicle, directly into the diesel exhaust gas stream where it breaks nitrogen oxide compounds (NOX) into its basic components of nitrogen and water. If the driver ignores DEF warning lights indicating that the DEF level is becoming critically low, the diesel's Powertrain Control Module (PCM) might limit vehicle speeds to about five miles per hour until the DEF level is restored.

7. Anti-Gelling Additives

In contrast to DEF, anti-gelling additives are added to the fuel tank to liquefy the solidified paraffin wax or "gel" that forms in the fuel and clogs diesel fuel filters during sub-freezing temperatures. While modern diesels use heated fuel lines and filters to prevent gelling, anti-gelling additives might be required to improve cold-starting performance during extreme winter temperatures. Most anti-gelling additives also include solvents that clean fuel systems, remove water and lubricate fuel injectors.

Characteristics and Properties of Fuels

1. Fuels are any material that store potential energy in forms, which upon burning in oxygen liberates heat energy.
2. Calorific value of fuel is the total quantity of heat liberated when a unit mass or volume of fuel is completely burnt.
3. Higher or gross calorific value (HCV) in the total amount of heat produced when a unit mass/volume of fuel has been burnt completely and the products of combustion have been cooled to room temperature (15°C or 60°F).
4. Lower or net calorific value (LCV) is the heat produced when unit mass (volume) of the fuel is burnt completely and the products are permitted to escape.
5. LCV = HCV – Latent heat of water formed.
6. Natural or primary fuels are found in nature such as wood, peat, coal, natural gas, petroleum.
7. Artificial or secondary fuels are prepared from primary fuels charcoal, coal gas, coke, kerosene oil, diesel oil, petrol, etc.
8. Fuels are further classified as,
   i. Solid Fuels.
   ii. Liquid Fuels.
   iii. Gaseous Fuels.

Characteristics of Solid fuels

1. Ash is high.
2. Low thermal efficiency.
3. Form clinker.
4. Low calorific value and require large excess air.
5. Cost of handling is high.
6. Cannot be used in IC engines.

Characteristics of Liquid Fuels

1. High calorific value.
2. No dust ash and clinker.

3. Clean fuels.
4. Less furnace air.
5. Less furnace space.
6. Used in IC engines.

Characteristics of Gaseous fuels

1. Have high heat content.
2. No ash or smoke.
3. Very large storage tanks are required.

An ideal Fuel should have the Following Properties

1. High calorific value.
2. Moderate ignition temperature.
3. Low moisture content.
4. Low NO_n combustible matter.
5. Moderate velocity of combustion.
6. Products of combustion not harmful.
7. Low cost.
8. Easy to transport.
9. Combustion should be controllable.
10. No spontaneous combustion.
11. Low storage cost.
12. Should burn in air with efficiency.

1.8.2 Potential and Advantages of Alternative Liquid and Gaseous Fuels for SI and CI Engines

Benefits/Advantages of Bio diesel

1. Bio diesel is bio renewable. Feed stocks can be renewed one or more times in a generation.
2. Bio diesel is carbon neutral. Plants use the same amount of CO2 to make the oil that is released when the fuel is burned.

3. Bio diesel is rapidly biodegradable and completely nontoxic, meaning spillages represent far less risk than petroleum diesel spillages.

4. Bio diesel has a higher flash point than petroleum diesel, making it safer in the event of a crash.

5. Blends of 20% bio diesel with 80% petroleum diesel can be used in unmodified diesel engines. Bio diesel can be used in its pure form but may require certain engine modifications to avoid maintenance and performance problems.

6. Bio diesel can be made from recycled vegetable and animal oils or fats.

7. Bio diesel is nontoxic and biodegradable. It reduces the emission of harmful pollutants, mainly particulates, from diesel engines (80% less CO_2 emissions, 100% less sulfur dioxide). But emissions of nitrogen oxide, the precursor of ozone, are increased.

8. Bio diesel has a high cetane number of above 100, compared to only 40% for petroleum diesel fuel. The cetane number is a measure of a fuel's ignition quality. The high cetane numbers of bio diesel contribute to easy cold starting and low idle noise.

9. The use of bio diesel can extend the life of diesel engines because it is more lubricating and, furthermore, power output is relatively unaffected by bio diesel.

Liquefied Natural Gas

Liquefied natural gas (LPG) is natural gas in its liquid form. When natural gas is cooled to -259F (161C), it becomes a clear, colorless, odorless liquid. LPG is generally refrigerated to 180°C for liquefaction and requires vacuum-insulated cryogenic tanks to maintain it in liquid form for storage. LPG is neither corrosive nor toxic.

Natural gas is primarily methane with low concentrations of other hydrocarbons, water, carbon dioxide, nitrogen, oxygen and some sulfur compounds. During the process known as liquefaction, natural gas is cooled below its boiling point, removing most of these compounds. The remaining natural gas is primarily methane with only small amounts of other hydrocarbons. Liquefying natural gas results in the purest form of methane when heated back to a gas.

For heavy-duty applications requiring long-range capability and large volumes of onboard fuel storage, LPG provides all the benefits of clean burning natural gas in a liquid form. LPG vehicles are essentially a heat exchanger installed between the fuel tank and the engine to warm the liquid and convert the fuel back to a gaseous state.

Advantages of Liquefied Natural Gas

1. LNG has very low particle emissions because of its low carbon to hydrogen ratio.
2. There are negligible evaporator emissions, requiring no relevant control.
3. Due to its low carbon-to-hydrogen ratio, it produces less carbon dioxide per GJ of fuel than diesel.
4. It has low cold-start emissions due to its gaseous state.
5. It has extended flammability limits, allowing stable combustion at leaner mixtures.
6. It has a lower adiabatic flame temperature than diesel, leading to lower NOx emissions.
7. LNG is pure methane which is a non-toxic gas.
8. It is much lighter than air and thus it is safer than spilled diesel.
9. Methane is not a volatile organic compound (VOC).

Disadvantages of Liquefied Natural Gas

1. There is considerable extra infrastructure involved with gas liquefaction.
2. It requires dedicated catalysts with high loading of active Catalytic components.
3. It is driving range is limited because its energy contained per value is relatively low.
4. Refueling is considered to be the least safe moment of its use.

1.9 Fuel Induction Techniques in SI and CI Engines

FIT

The fuel induction techniques (FIT) have been found to be playing a very dominant and sensitive role in determining the performance characteristics of an I.C. Engine.

The 'FIT' for an S.I. engine can be classified into four categories such as,

1. Carburetion.
2. Inlet Port Injection.

3. Direct Cylinder Injection.
4. Inlet Manifold Injection.

These conventional methods of 'FIT' could also be applied to engine operation with a non-conventional alternative fuel, such as hydrogen. These methods of carburetion by the use of a gas carburetor have been the very simplest and the oldest technique. In a gasoline-fueled engine, the volume occupied by the fuel is about 1.7% of the mixture whereas a carburetted hydrogen engine, using gaseous hydrogen, results in a power output loss of 15%.

Thus, from eliminating unwanted combustion symptoms, fuel induction techniques have also been quite effective in compensating for the power loss. The Injection of hydrogen into the inlet manifold offers an alternative to the conventional load control method by throttling. This method uses the typical properties of hydrogen fuel to a point of advantage. It also possesses the ability to initiate fuel delivery at a timing position sometime after the beginning of intake stroke.

The system could be so designed that the intake manifold does not contain any combustible mixture thereby avoiding extreme situation leading to undesirable combustion phenomena. In a carburetted engine system, the valve overlap between the exhaust and the intake stroke can bring the fuel-air charge into contact with the residual hot gases.

However, if by any chance pre-ignition does take place during intake stroke, it will have much lesser consequence as compared to that occurring in a carburetted engine. Some investigators have also carried out research on intake port injection. In such a system each air and fuel enters the combustion chamber during the intake stroke, but are not pre-mixed in the intake manifold.

Direct cylinder injection of hydrogen into the combustion chamber does have all the benefits of the late injection as characterized by manifold injection. In addition, the system permits for fuel delivery after the closure of the intake valve and thus, intrinsically precludes the possibility of backfire. The injection system will have to cater to some stringent requirements in respect of the severe thermal environment which the injector is bound to encounter.

All the mechanical parts which form part of the injection system must be able to withstand such a high pressure, say to the tune of about 100 atm. When considering a practical automobile, maintaining a maximum pressure such as about 100 bars, in a vehicle for onboard storage methods raises serious problems.

1.9.1 Mixture Requirements at Different Loads and Speeds

The air-fuel ratio at which an engine operates has a considerable influence on its performance. Consider an engine operating at full throttle and constant speed with varying

A/F ratio. Under these conditions, the A/F ratio will affect both the power output and the brake specific fuel consumption, as indicated by the typical curves shown in figure.

The mixture corresponding to the maximum output on the curve is called the best power mixture with an A/F ratio of approximately 12:1. The mixture corresponding to the minimum point on the curve is called the best economy mixture. The A/F ratio is approximately 16:1.

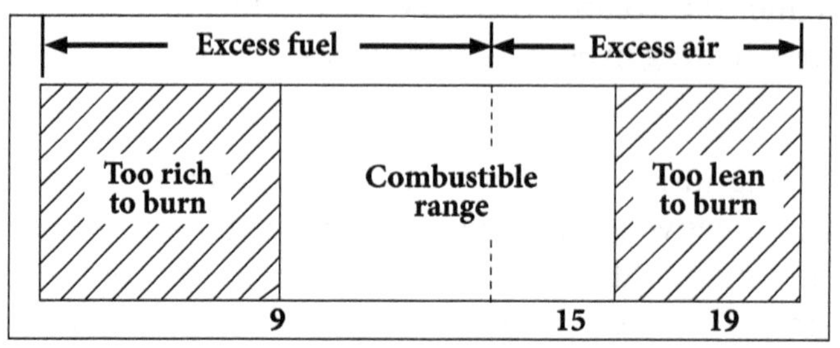

Air-fuel ratio Useful air-fuel mixture range of gasoline.

It may be noted that the best power mixture is much richer than the chemically correct mixture and the best economy mixture is slightly leaner than the chemically correct.

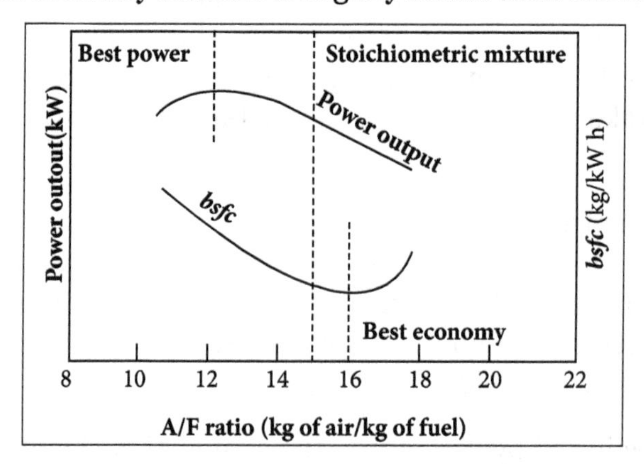

The figure above showing a variation of power output with air-fuel ratio for an SI engine is based on full throttle operation. The A/ F ratios for the best power and best economy at part throttle are not strictly the same as at full throttle. If the A/F ratios for best power and best economy are constant over the hill range of throttle operation and if the influence of other factors is disregarded, the ideal fuel metering device would be merely a two position carburetor.

Such a carburetor could be set for the best power mixture when maximum performance is desired and for the best economy mixture when the primary consideration is the fuel economy. These two settings are indicated in the figure by the solid horizontal lines X-X' and Z-2', respectively. Actual engine requirements, however, again preclude the use of such a simple and convenient arrangement. These requirements are discussed in the succeeding section.

Under normal conditions it is desirable to run the engine on the maximum economy mixture, viz., around 16:1 air-fuel ratio. For quick acceleration and for maximum power, rich mixture viz., 12:1 air-fuel ratio is required.

1.10 Carburetion: Factors Affecting Carburetion

The main components of simple carburetor are: Float chamber, float, nozzle, venture, inlet valve and metering jet. In the float chamber, a constant level of petrol is maintained by the float and a needle valve. The float chamber is ventilated to atmosphere. This is used to maintain atmospheric pressure inside the chamber.

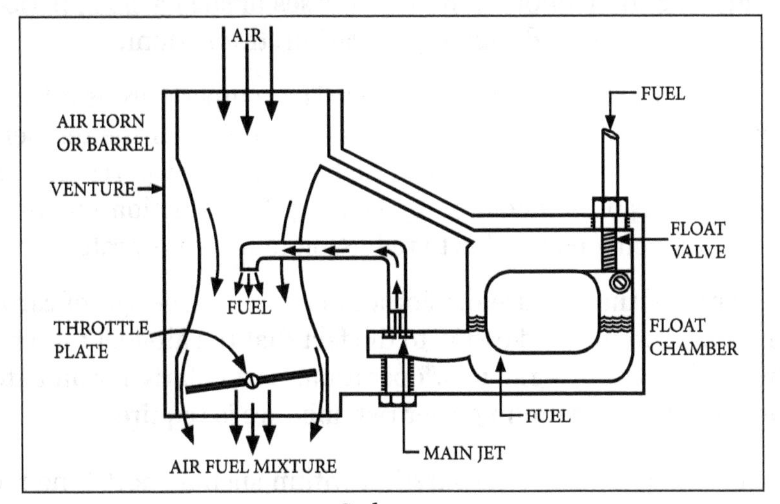

Carburetor

The float which is normally a metallic hollow cylinder rise and close the inlet valve as the fuel level in the float chamber increases to certain level.

The mixing chamber contains venture, nozzle and throttle valve, the venturi tube is fitted with the inlet manifold. This tube is narrow opening called venture. A nozzle provides jet below the center of this venture. The nozzle keeps the same level of petrol as the level in the float chamber. The mixing chamber has two butterfly valves. One is to allow air in to the mixing chamber known as choke valve. The other is to allow air fuel mixture to the engine known as throttle valve.

Working

During the suction stroke, vacuum is created inside the cylinder. This causes pressure difference between the cylinder and outside the carburetor. Due to this, the atmospheric air enters in to the carburetor. The air flows through the venturi. This produces the partial vacuum at the tip of the nozzle. Because of this vacuum, the fuel comes out from the nozzle in the form of fine spray. These fine fuels partially mix with incoming air to form air-fuel mixture. Thus, it gives a homogeneous mixture of air fuel to the engine.

The various factors which influence the process of carburetion are as follows:

1. The engine speed: The time available for the preparation of the mixture. In case of modern high speed engines, the time duration available for the formation of mixture is very small and limited. The time duration for mixture formation and induction may be of the order of 10 to 5 milliseconds.

2. The vaporization characteristics of fuel: Atomisation, mixing and vaporization are the processes which require a finite time to occur. The time available for mixture formation is very small in high speed engines (For example, in an engine running at 3000 r.p.m., the induction process lasts for less than 0.02 second). For completion of these processes in such a small period a great ingenuity is required in designing the carburettor system.

3. The temperature of the incoming air: The temperature is a factor which effectively controls vaporization process of the fuel. If the temperature of the incoming air is high, it results in higher rates of vaporization. The mixture temperature can be increased by heating the induction manifold but it will result in reducing power due to reduction in mass flow rate.

4. The design of the carburetor: For a SI engine, the design of carburetion system is very complicated owing to the fact that the air-fuel ratio required by it varies widely over its range of operation, particularly for an automotive engine. For idling maximum power rich mixture is required.

   To achieve high quality carburetion within such a short time requires good vaporization characteristics of the fuel which are ensured by presence of high volatile hydrocarbons in the fuel.

1.10.1 Principle of Carburetion

Both air and gasoline are drawn through the carburetor and into the engine cylinders by the suction created by the downward movement of the piston. This suction is due to an increase in the volume of the cylinder and a consequent decrease in the gas pressure in this chamber. It is the difference in pressure between the atmosphere and cylinder that causes the air to flow into the chamber.

In the carburetor, air passing into the combustion chamber picks up fuel discharged from a tube. This tube has a fine orifice called carburetor jet which is exposed to the air path. The rate at which fuel is discharged into the air depends on the pressure difference or pressure head between the float chamber and the throat of the venturi and on the area of the outlet of the tube.

In order that the fuel drawn from the nozzle may be thoroughly atomized; the suction effect must be strong and the nozzle outlet comparatively small. In order to produce a strong suction, the pipe in the carburetor carrying air to the engine is made to have a restriction.

At this restriction called throat due to increase in velocity of flow, a suction effect is created. The restriction is made in the form of a venturi as shown in figure to minimize throttling losses. The end of the fuel jet is located at the venturi or throat of the carburetor.

The geometric venturi tube is shown in figure given below. It has a narrower path at the centre so that the flow area through which the air must pass is considerably reduced. As the same amount of air must pass through every point in the tube, its velocity will be greatest at the narrowest point. The smaller the area, the greater will be the velocity of the air and thereby the suction is proportionately increased.

In order that the fuel drawn from the nozzle may be thoroughly atomized; the suction effect must be strong and the nozzle outlet comparatively small. In order to produce a strong suction, the pipe in the carburetor carrying air to the engine is made to have a restriction.

At this restriction called throat due to increase in velocity of flow, a suction effect is created. The restriction is made in the form of a venturi as shown in figure to minimize throttling losses. The end of the fuel jet is located at the venturi or throat of the carburetor.

The geometric venturi tube is shown in figure given below. It has a narrower path at the centre so that the flow area through which the air must pass is considerably reduced. As the same amount of air must pass through every point in the tube, its velocity will be greatest at the narrowest point. The smaller the area, the greater will be the velocity of the air and thereby the suction is proportionately increased.

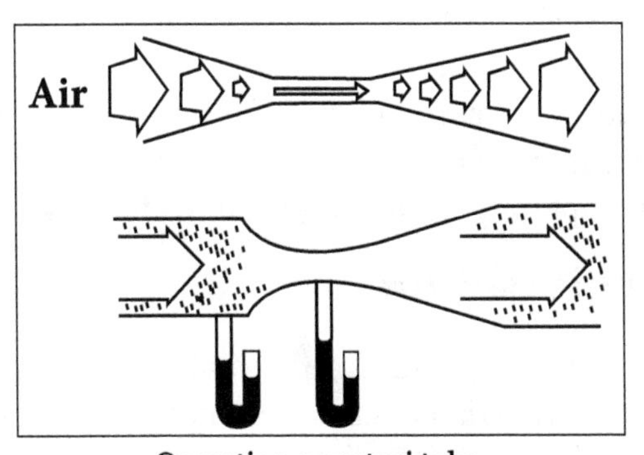

Operation o venturi tube.

As mentioned earlier, the opening of the fuel discharge jet is usually located where the suction is maximum. Normally, this is just below the narrowest section of the venturi tube. The spray of gasoline from the nozzle and the air entering through the venturi tube are mixed together in this region and a combustible mixture is formed which passes through the intake manifold into the cylinders. Most of the fuel gets atomized and simultaneously a small part will be vaporized. Increased air velocity at the throat of the venturi helps the rate of evaporation of fuel.

The difficulty of obtaining a mixture of sufficiently high fuel vapour-air ratio for efficient starting of the engine and for uniform fuel-air ratio in different cylinders (in case of multi-cylinder engine) cannot be fully met by the increased air velocity alone at the venturi throat.

1.10.2 Simple Carburetor and its Drawbacks

The function of a Carburettor is to vaporize the petrol (gasoline) by means of engine suction and to supply the required air and fuel (petrol) mixture to the engine cylinder. During the suction stroke, air flows from atmosphere into the cylinder. As the air passes through the venture, velocity of air increases and its pressure falls below the atmosphere. The pressure at the nozzle tip is also below the atmospheric pressure.

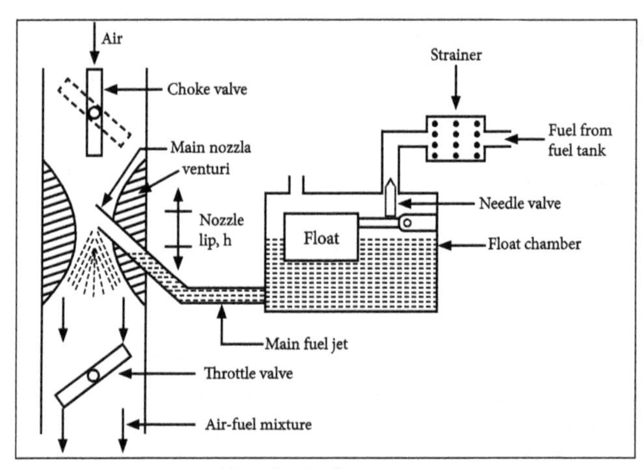

Simple Carburettor.

The pressure on the fuel surface of the fuel tank is atmospheric due to which a pressure difference is created, which causes the flow of fuel through the fuel jet into the air stream. As the fuel and air pass ahead of the venturi, the fuel gets vaporized and required uniform mixture is supplied to the engine. The quantity of fuel supplied to the engine depends upon the opening of throttle valve which is governed by the governor.

Drawbacks of a Simple Carburettor

1. In Simple Carburettor, the mixture is weakened when the throttle is suddenly opened because of Inertia effect of the Fuel which prevents the proper quantity of Fuel from flowing immediately.

2. The working of Simple Carburettor is affected by changes of Atmospheric Pressure. Carburettor used in Aircraft is to be provided with Altitude control, as the rich mixture is unnecessarily available, due to less Density of Air.

3. The working of Simple Carburettor is affected by changes of Atmospheric Temperature. If the setting is done in Winter season, it will be found to give too rich mixture in the Summer. This is due to less density of air with the rise of temperature to a greater extent than the density of fuel.

4. At a very low speed, the mixture supplied by a Simple Carburettor is so weak that it will not ignite properly and for its enrichment, at such conditions some arrangement in the Carburettor is required to be made.

5. It gives the proper mixture at only one engine speed and load, therefore, suitable only for Engines running at constant speed, the quantity of Fuel issuing out will change and not match the Velocity of Air flowing through the Venturi. To overcome this various modifications have to be made in Simple Carburettor.

1.10.3 Calculation of the Air–Fuel Ratio

A simple carburetor with the tip of the fuel nozzle h meters above the fuel level in the float chamber is shown in figure below. It may be noted that the density of air is not the same at the inlet to the carburetor (section A-A, point 1) and the venturi throat (section B-B, point 2). The calculation of exact air mass flow involves taking this change in density or compressibility of air into account.

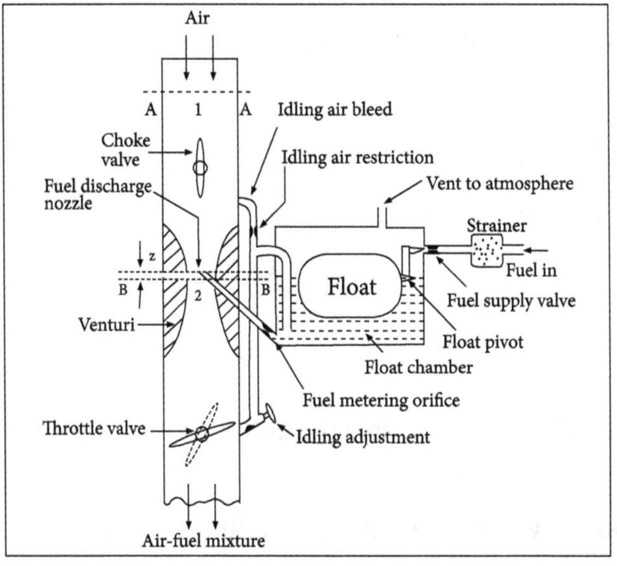

Carburetor

Applying the steady flow energy equation to sections AA and B-B and assuming unit mass flow of air, We have,

$$q - w = (h_2 - h_1) + \frac{1}{2}(C_2^2 - C_1^2) \qquad ...(1)$$

Here q, w are the heat and work transfers from entrance to throat and h and C stand for enthalpy and velocity respectively.

Assuming an adiabatic flow, we get q=0, w=0 and $C_1 \approx 0$

$$C_2 = \sqrt{2(h_1 - h_2)} \qquad ...(2)$$

Assuming air to behave like ideal gas, we get $h = C_p T$. Hence, Equation 2 can be written as,

$$C_2 = \sqrt{2C_p(T_1 - T_2)} \quad ...(3)$$

As the flow process from inlet to the venturi throat can be considered to be isentropic. We have,

$$\frac{T_2}{T_1} = \left(\frac{p_2}{p_1}\right)^{\left(\frac{\gamma-1}{\gamma}\right)} \quad ...(4)$$

$$T_1 - T_2 = T1\left[1 - \left(\frac{p_2}{p_1}\right)^{\left(\frac{\gamma-1}{\gamma}\right)}\right] \quad ...(5)$$

Substituting equation 5 in equation 3, we get

$$C_2 = \sqrt{2C_p T_1\left[1 - \left(\frac{p_2}{p_1}\right)^{\left(\frac{\gamma-1}{\gamma}\right)}\right]} \quad ...(6)$$

Now, mass flow of air,

$$m_a = \rho_1 A_1 C_1 = \rho_2 A_2 C_2 \quad ...(7)$$

Where A_1 and A_2 are the cross-sectional area at the air inlet (point 1) and venturi throat (point 2).

To calculate the mass flow rate of air at venturi throat, we have:

$$p_1 / \rho_1^{\gamma} = p_2 / \rho_2^{\gamma} \quad ...(8)$$

$$\rho_2 = (p_2 / p_1)^{1/\gamma} \rho_1$$

$$\dot{m}_a = \left(\frac{p_2}{p_1}\right)^{1/\gamma} \rho_1 A_2 \sqrt{2C_p T_1\left[\left(1 - \frac{p_2}{p_1}\right)^{\frac{\gamma-1}{\gamma}}\right]}$$

$$= \left(\frac{p_2}{p_1}\right)^{1/\gamma} \frac{p_1}{RT_1} A_2 \sqrt{2C_p T_1\left[\left(1 - \frac{p_2}{p_1}\right)^{\frac{\gamma-1}{\gamma}}\right]} \quad ...(9)$$

$$= \frac{A_2 p_1}{R\sqrt{T_1}} \sqrt{2C_p \left[\left(\frac{p_2}{p_1} \right)^{\frac{2}{\gamma}} - \left(\frac{p_2}{p_1} \right)^{\frac{\gamma+1}{\gamma}} \right]} \qquad ...(10)$$

Substituting Cp =1005 J/kg K, γ =1.4 and R=287 J/kg K for air,

$$\dot{m}_a = 0.1562 \frac{A_2 p_1}{\sqrt{T_1}} \sqrt{\left(\frac{p_2}{p_1} \right)^{1.43} - \left(\frac{p_2}{p_1} \right)^{1.71}}$$

$$= 0.1562 \frac{A_2 p_1}{\sqrt{T_1}} \phi \text{ kg/s} \qquad ...(11)$$

Where,

$$\phi = \sqrt{\left(\frac{p_2}{p_1} \right)^{1.43} - \left(\frac{p_2}{p_1} \right)^{1.71}} \qquad ...(12)$$

Here, p is in N/m^2, a is in m^2 T is in K.

Equation 11 gives the theoretical mass flow rate. To get the actual mass flow rate, the above equation should be multiplied by the co-efficient of discharge for the venturi, C_{da}.

$$\dot{m}_{a_{actual}} = 0.1562 C_{da} \frac{A_2 p_1}{\sqrt{T_1}} \phi \qquad ...(13)$$

Since C_{da} and A_2 are constants for a given venturi,

$$\dot{m}_{a_{actual}} \propto \frac{p1}{\sqrt{T_1}} \phi \qquad ...(14)$$

In order to calculate the air-fuel ratio, fuel flow rate is to be calculated. As the fuel is incompressible, applying Bernoulli's theorem we get,

$$\frac{p_1}{\rho_f} - \frac{p_2}{\rho_f} = \frac{C_f^2}{2} + gz \qquad ...(15)$$

Where ρ_f is the density of fuel, C_f is the fuel velocity at the nozzle exit and z is the height of the nozzle exit above the level of fuel flow,

$$C_f = \sqrt{2 \left[\frac{p_1 - p_2}{\rho_f} - gz \right]}$$

Mass flow rate of fuel,

$$\dot{m}_f = A_f C_f \rho_f \qquad ...(16)$$

$$= A_f \sqrt{2\rho_f (p_1 - p_2 - gz\rho_f)} \qquad ...(17)$$

Where A_f is the area of cross-section of the nozzle and ρ_f is the density of the fuel,

$$\dot{m}_{f_{actual}} = C_{df} A_f \sqrt{2\rho_f (p_1 - p_2 - gz\rho_f)} \qquad ...(18)$$

Where C_{df} is the coefficient of discharge for fuel nozzle,

$$A/F \text{ ratio} = \frac{\dot{m}_{a_{actual}}}{\dot{m}_{f_{actual}}}$$

$$\frac{A}{F} = 0.1562 \frac{C_{da}}{C_{df}} \frac{A_2}{A_f} \frac{p_1 \phi}{\sqrt{2T_{1pf} (p_1 - p_2 - gz\rho_f)}} \qquad ...(19)$$

1.10.4 Modern Carburetors

Apart from the above compensating devices there are few other systems normally used in modern carburetors for meeting the requirement of vehicles. The details of the various systems are explained in the following sections.

Anti-Dieseling System

An SI engine sometimes continues to run for a very small period even after the ignition is switched off. This phenomenon is called dieseling (after running or run-on).

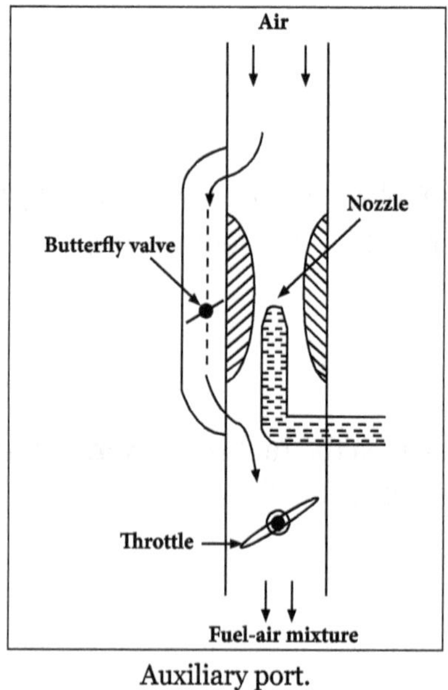

Auxiliary port.

Dieseling may take place due to one or more of the following:

i. Engine idling speed is set to high.

ii. Increase in compression ratio due to carbon deposits.

iii. Inadequate or low octane rating.

iv. Engine overheating.

v. Too high spark plug heat range.

vi. Incorrect adjustment of idle fuel-air mixture (usually toluene).

vii. Sticking of throttle.

viii. Requirement of tune up of engine.

ix. Oil entry into the cylinder.

Some modern automobiles use anti dieseling system to prevent dieseling system. This system has a solenoid valve operated idling circuit. With ignition key turned on current flows in the solenoid coil of the solenoid valve generating a force. This force pulls the needle valve and opens the passage for slow mixture. When the ignition key is turned off the magnetic force disappears.

Then the needle valve is brought to the original by the action of the spring in the solenoid valve. By this way the slow mixture passage is cut-off and hence the engine stops. This reduces hydrocarbon emissions.

Richer Coasting System

The richer coasting system is incorporated in some modern cars. When the car is traveling at high speed and when the accelerator pedal is suddenly released, the wheel will motor the engine at a high speed. Consequently, the vacuum in the inlet manifold and the combustion chamber increases too much and causes incomplete combustion.

The richer coaster system is designed to overcome this problem by supplying a proper mixture to the intake manifold for proper combustion. This system has a chamber connected to the intake manifold for stable combustion. When the throttle valve is closed to decelerate, the vacuum of the intake manifold increase.

As this happens the vacuum applied to the chamber pulls the membrane and cause the coasting valve to open. Then the fuel in the float chamber is metered at the coasting fuel jet and mixed with air and sucked into the intake manifold.

Acceleration Pump System

Acceleration is a transient phenomenon. In order to accelerate the vehicle and consequently its engine, the mixture required is very rich and the richness of the mixture has to be obtained quickly and very rapidly. In automobile engines situations arise when it is necessary to accelerate the vehicle. This requires an increased output from the engine in a very short time. If the throttle is suddenly opened there is a corresponding increase in the air flow.

However, because of the inertia of the liquid fuel, the fuel flow does not increase in proportion to the increase in air flow. This results in a temporary lean mixture causing the engine to misfire and a temporary reduction in power output. To prevent this condition, all modern carburetors are equipped with an accelerating system. The plunger moves into the cylinder and forces an additional jet of fuel at the venturi throat. When the throttle is partly open, the spring sets the plunger back. There is also an arrangement which ensures that fuel in the pump cylinder is not forced through the jet when valve is slowly opened or leaks past the plunger or some hole into the float chamber.

Mechanical linkage system, in some carburetor, is substituted by an arrangement whereby the pump plunger is held up by manifold vacuum. When this vacuum is decreased by rapid opening of the throttle, a spring forces the plunger down pumping the fuel through the jet.

Economizer or Power

Enrichment System At the maximum power range of operation from 80% to 100% load, richer air-fuel ratio of about 12 to 14 is required and the maximum power, an air-fuel ratio of approximately 12 is expected. An economizer is a valve which remains closed at normal cruise operation and gets opened to supply rich mixture at full throttle operation. It regulates the additional fuel supply during the full throttle operation.

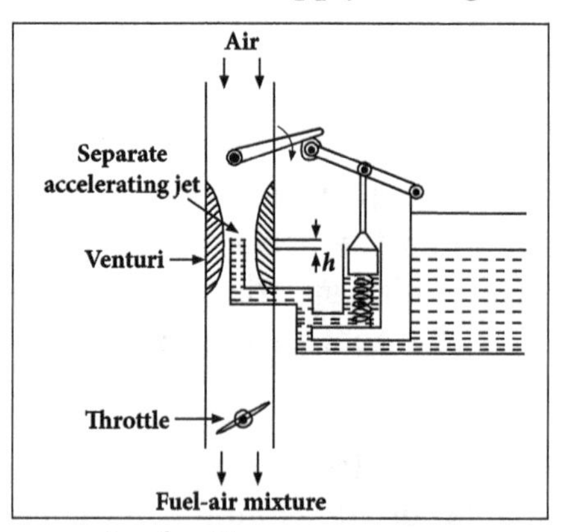

Acceleration pump system.

The term economizer is rather misleading. Probably as it does not interfere during cruising operation where an economy mixture is supplied it is called economizer. It should more appropriately be called power enrichment system. Above figure shows the skeleton outline of a metering rod economizer system. It allows a large opening to the main jet only when the throttle is opened beyond a specified limit. The metering rod may be tapered or stepped.

2

Fuel Injection, Ignition, Combustion, Super Charging and Scavenging

2.1 Fuel Injection: Functional Requirements of an Injection System

1. Introduction of the fuel into the combustion chamber should take place within a precisely defined period of the cycle.
2. The metering of the amount of fuel injected per cycle should be done very accurately.
3. The quantities of the meter should vary to meet the changing load and speed requirements.
4. The injection rate should be such that it results in the desired beat release pattern.
5. The injected fuel must be broken into very fine droplets.
6. The pattern of spray should be such as to ensure rapid mixing of fuel and air.
7. The beginning and end of the injection should be sharp.
8. The timing of injection, if desired, should change as per the requirements of load and speed.
9. The distribution of the metered fuel, in the case of multi-cylinder engines, should be uniform among various cylinders.
10. Besides above requirements, the weight and the size of the fuel injection system must be minimum. It should be cheaper to manufacture and least expensive to attend to adjust or repair.

For accomplishing these requirements the following functional elements are required in a fuel injection system:

1. Pumping elements- To move the fuel from the fuel tank to cylinder and piping etc.
2. Metering elements- To measure and supply the fuel at the rate demanded by the load and speed.

3. Metering controls- To adjust the rate of metering elements for changes in load and speed of the engine.

4. Distributing elements- To divide the metered fuel equally among the cylinders.

5. Timing controls- To adjust the start and the stop of injection.

6. Mixing elements- To atomise and distribute the fuel within the combustion chamber.

The main functions of a fuel injection system are:

1. Filter the fuel.
2. Metre or measure the correct quantity of fuel to be injected.
3. Time the fuel injection.
4. Control the rate of fuel injection.
5. Atomise or break up the fuel to fine particles.
6. Properly distribute the fuel in the combustion chamber.

The injection systems are manufactured with great accuracy, especially the parts that actually metre and inject the fuel. Some of the tolerances between the moving parts are very small of the order of one micron. Such closely fitting parts require special attention during manufacture and hence the injection systems are costly.

2.2 Classification of Injection Systems

Thus the injection systems can be classified as:

1. Air fuel injection system.
2. Solid injection systems.

Air Injection

In this method of fuel injection air is compressed in the compressor to a very high pressure (much higher than developed in the engine cylinder at the end of the compression stroke) and then injected through the fuel, mule into the engine cylinder. The rote of fuel admission can be controlled by varying the pressure of injection air. Storage air bottles which are kept charged by an air compressor (driven by the engine) supply the high pressure air.

Advantages

i. It provides better automatization and distribution of fuel.

ii. As the combustion is more complete, the b.m.e.p is higher than with other types of injection system.

iii. Inferior fuels can be used.

Disadvantages

i. It requires a high pressure multi-stage compression. The large number of parts, the intercooler etc. make the system complicated and expensive.

ii. A separate mechanical linkage is required to time the operation of fuel valve.

iii. Due to the compression and the linkage the bulk of the engine increases. This also results in reduced B.P. due to power less in operating the compression and linkage.

iv. The fuel in the combustion chamber bums very near to injection nozzle which many times leads to overheating end burning of valve and its seat.

v. The fuel valve sealing requires considerable skill.

vi. In case of sticking of fuel valve the system becomes quite dangerous due to the presence of high Pressure air.

Solid or Airless Injection

Injection of fuel directly into the combustion chamber without primary automation is termed as solid injection. It is also termed as mechanical injection.

Main Components

The main components of a fuel injection system are:

i. Fuel tank.

ii. Fuel feed pump to supply the fuel from the main fuel tank to the injection pump.

iii. Fuel filters to prevent dust and abrasive particles from entering the pump and injectors.

iv. Injection pumps to meter and pressurize the fuel for injection.

v. Governor to ensure that the amount of fuel is in accordance with variation in load.

vi. Fuel piping's and injectors to take the fuel from the pump and distribute it in the combustion chamber by atomizing it in fine droplets.

Main types of modern fuel injection systems:

1. Common-rail injection system.
2. Individual pump injection system.
3. Distributor system.

Atomisation of fuel oil has been secured by (i) air blast and (ii) pressure spray. Equally diesel engines use air fuel injection at about 70 bar. This is sufficient not only to inject the oil, but also to atomise it for a rapid end through combustion. The expense of providing an air compressor and tank lead to the development or solid injection, using a liquid pressure of between 100 and 200 bar which is sufficiently high to atomise the oil it forces through spray nozzles.

Great advances have been made in the field of solid injection of the fuel through research and progress in fuel pump, spray nozzles and combustion chamber design.

1. Common-Rail Injection System

Two types of common-rail injection systems are shown in figure (a) and (b) respectively.

i. A single pump supplies high-pressure fuel to header, a relief valve holds pressure constant. The control wedge adjusts the lift of mechanical operated valve to set amount and time of injection.

ii. Controlled-pressure system has pump which maintains head pressure. Pressure relief and timing valves regulate injection time and mount. Spring loaded spray valve acts merely as a check.

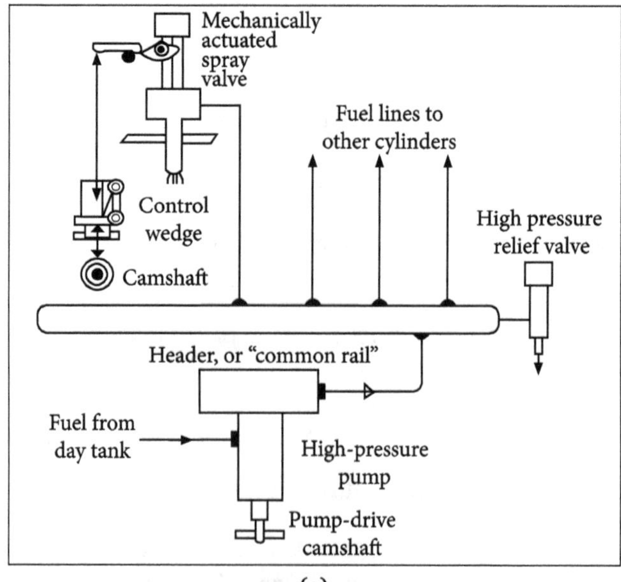

(a)

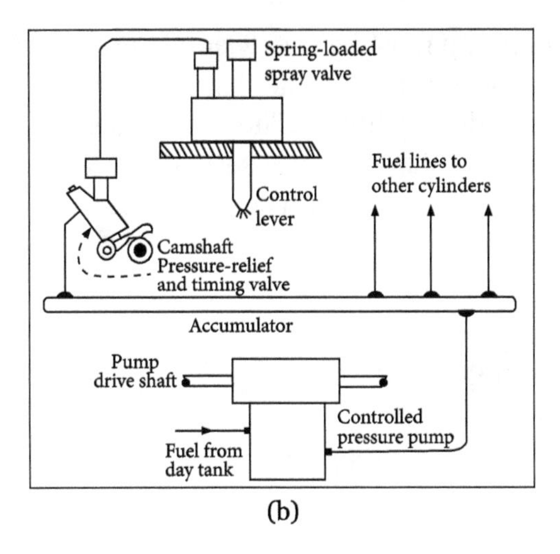

(b)

Advantages

i. The system arrangement is simple and less maintenance coat.

ii. Only one pump is sufficient for multi-cylinder engine.

iii. It fulfills the requirements of either the constant load with variable speed or constant speed with variable load.

iv. Variation in pump supply pressure will affect the cylinders uniformly.

Disadvantages

i. There is a tendency to develop leaks in the injection valve.

ii. Very accurate design and workmanship are required.

2. Individual Pump Dejection System

In this system on individual pump or pump cylinder connects directly to each fuel nozzle. Pump meters charge and control injection timing. Nozzles contain a delivery valve actuated by the fuel-oil pressure.

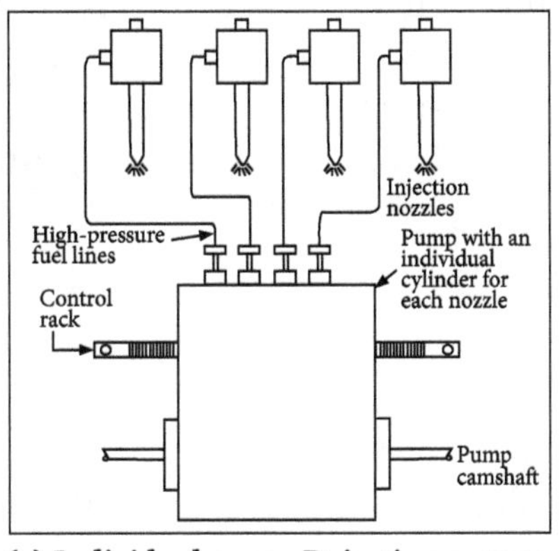

(c) Individual pump Dejection system.

The design of this type of pump must be very accurate and precise as the volume of fuel injected per cycle is 1/20,000 of the engine displacement at full load and 1/100,000 of the engine displacement during Idling. The time allowed for injecting such a small quantity of fuel is very limited (about 1/460 second at 1600 r.p.m. of the engine providing injection through 20° crank angle). The pressure requirements vary from 100 to 300 bar.

3. Distributer System

In this system, the fuel is metered at a central point a pump pressurize, meters the fuel and times the injection.

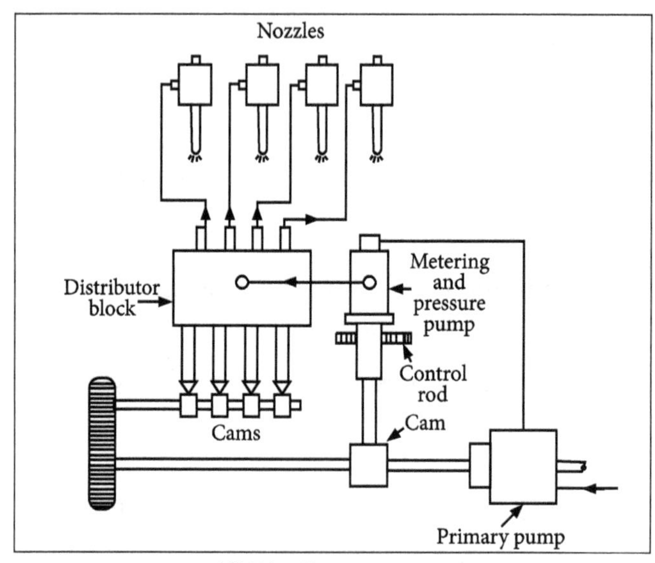

(d) Distributer system.

2.2.1 Fuel Feed Pump and Injection Pump

Fuel Feed Pumps

The fuel feed pump used for the diesel engine is similar to that of a fuel lift pump for the petrol engine. It delivers the fuel from the tank to the injection pump continuously and at a reasonable pressure. It is necessary because there is possibility of formation of vapour bubbles and subsequently cavitation in the pump due to suction of the rapidly moving plungers of the injection pump. This would lead to uncontrolled variations in the rate of delivery of fuel to the cylinders, causing rough running and possibly even mechanical damage to the engine.

Also cavitation could cause mechanical damage in the injection pump. Commonly delivery pressures of between about 29 and 98 kPa is adequate for preventing vapour formation on the suction side of in-line type injection pumps. This pressure also ensures adequate supply of fuel for filling the plunger elements at high speeds in a rotary or distribution pumps.

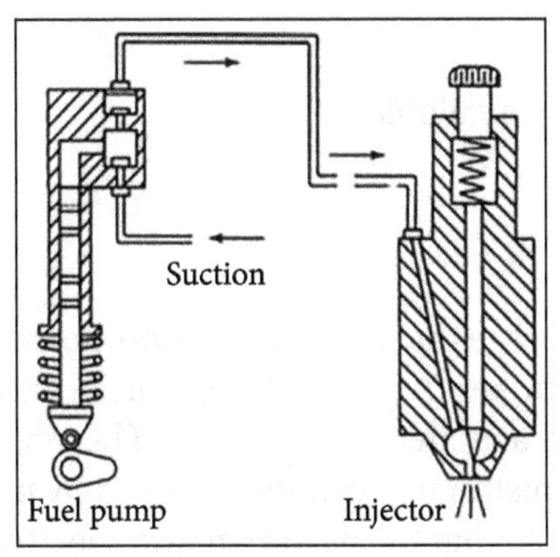

Fuel Feed Pumps.

Diaphragm Type Pumps

The Diaphragm kind of fuel transfer pumps are similar to the lift pumps used for petrol engines. Also the basic principle of operation of the AC pump shown in figure is similar to the above, except that it is designed for direct cam actuation, without a lever type follower. In this case the rod connected to the diaphragm is pulled up instead of pulled down by the cam mechanism.

The return spring used for the diaphragm is weaker than that used for the push rod. Consequently, as the cam nose rotates past its top dead centre, the push-rod is pulled down by its return spring until the head formed at its upper end bears against the seating in the boss of the diaphragm carrier, so that the diaphragm is pulled down with it.

The pressure differential in the chamber below the diaphragm causes the inlet valve to open so that it is filled with fuel. On the return stroke, the inlet valve closes and the delivery valve opens and the fuel is delivered by the action of the diaphragm return spring, however only at the rate to satisfy the demands of the engine.

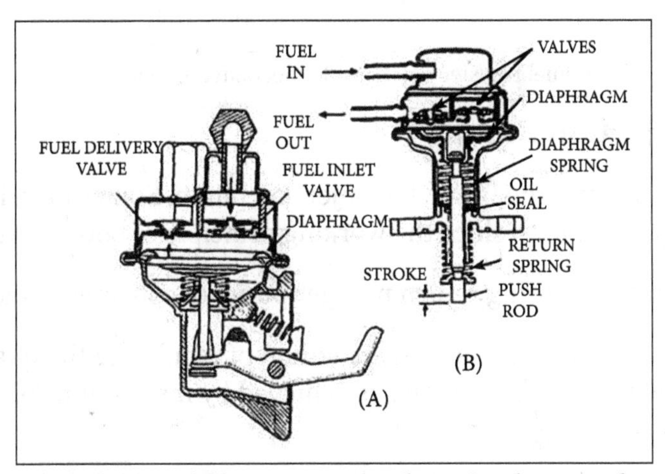

A. Unit AC model lever arm actuated. B. Directly actuated.

Plunger Type Pumps

Two plunger type pumps are available;

i. Single acting.

ii. Double acting.

The former has the disadvantage that in extreme conditions its delivery pressure can fall to zero between pumping strokes. In some models these pumps can be flange-mounted on the injection pumps and actuated by the cams. The single-acting version shown in figure uses a cam-actuated piston in a cylinder flanked by non-return valve. The piston is lifted by its return spring, so that fuel is drawn into the pressure chamber through the port on the right and a filter and non-return valve.

When the cam, acting on a roller-follower and push rod, forces the piston down again, the non-return valve on the inlet side closes and that over the transfer port on the other side opens, allowing the fuel to pass up into the chamber above the piston. As the cam follower and the push rod rise, the piston is pushed upwards by its return spring generating pressure, which closes the non-return valve over the transfer port, displacing the fuel above it through the outlet port to the injection pump at the rate required by the engine. The double-acting version has no transfer valve, but has two inlet and two delivery valves. With both stroke, fuel is drawn directly through the filter into the chamber on one side of the piston while, that on the other side it is delivered to the injection pump.

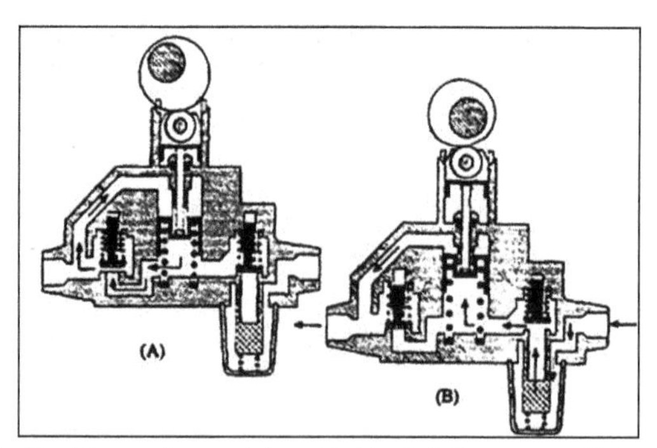

A. Fuel transfer position. B. Fuel delivery condition.

Bosch Fuel Injection Pump

i. Figure (a) shows the Bosch fuel injection pump. Sometimes, it is also called as Helix by bass pump since the working principle of both is same.

ii. This pump has spring return plunger pump of constant stroke type.

iii. Its special features are effective method of regulating the quality of fuel to be delivered to the cylinder. This is achieved by means of helical grooved cut on the plunger.

iv. Spill ports always remain full of fuel since it is connected to the fuel line.

v. Each pump consists of a steel pump barrel which fits accurately in a steel pump plunger.

vi. A spring loaded delivery valve is fitted at the upper end of the barrel.

Various positions of plunger are shown in figure (b).

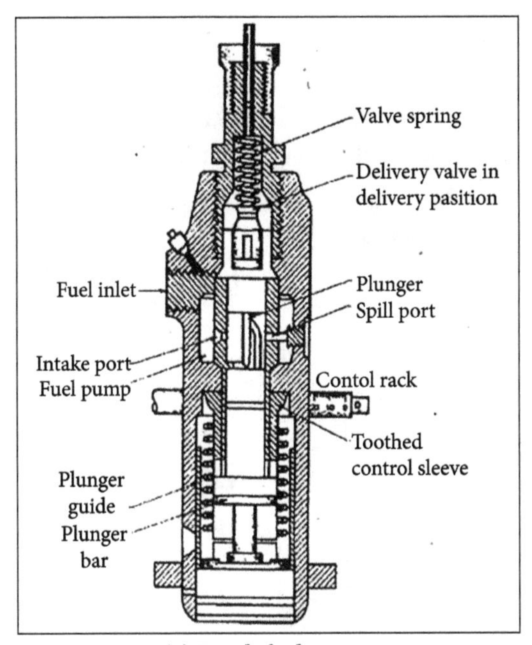

(a) Bosch fuel pump.

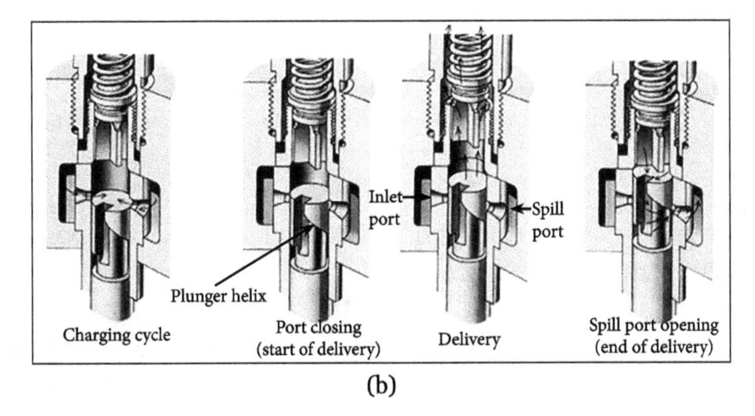

(b)

i. Fuel flows by gravity from a fuel sump in the pump unit body. During the downward motion of the plunger, a partial vacuum is created and as a result the fuel flows from the sump through the intake port (a) into the barrel as shown in figure (b and a).

ii. The plunger moves vertically in its barrel with a constant stroke. During the upward stroke of the plunger, it covers the intake port and compresses the fuel which flows past the delivery valve.

Jerk Type Diesel Fuel Injection Pump: CAV Fuel Injection Pump

The function of fuel injection pump is to measure and deliver the correct quantity of fuel at high pressure to the injector. CAV fuel injection pump is most commonly used in diesel engines. Figure, shows a line diagram of CAV fuel injection pump.

Construction

It is a plunger type pump. It has a reciprocating plunger which reciprocates inside the barrel. The plunger has a helical groove. This groove extends as vertical groove to the top. The plunger is operated by a cam mechanism.

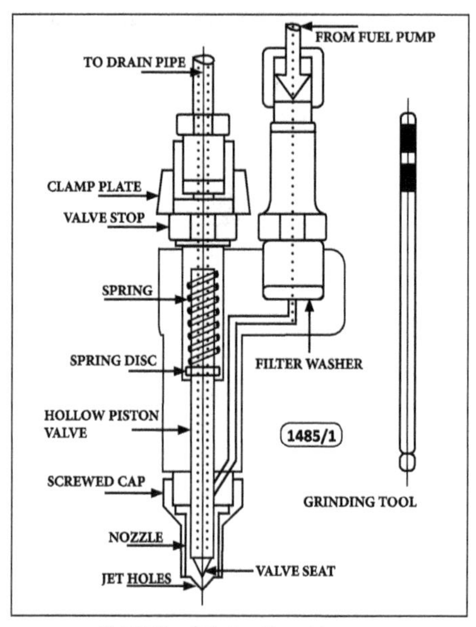

CAV Fuel Injection Pump.

Working

The plunger is moved up by a cam and return back to its initial position by tension spring. The plunger can also be rotated by the rack and pinion arrangement. Fuel delivery valve is seated in its seat by the force of the spring. Two ports are provided in the barrel. One is known as supply port and the other is known as spill port. This port is opened and closed by the moving plunger. The fuel passage is connected to fuel injector.

When the supply port is opened, the fuel is filled in the barrel. When the cam rotates, it will lift the plunger up. The moving plunger first closes the supply and the spill ports. Then the fuel above the plunger is compressed and high pressure is developed. Due to this high pressure, the delivery valve is lifted off and the fuel flows through the fuel passage to the atomizer.

The plunger moves further, the top portion of plunger is connected to the spill port by helical groove. The remaining fuel in the barrel comes out through the helical groove when the plunger moves up. So, the fuel pressure falls in the barrel. Due to this, the delivery valve is brought back to its seat by the spring force. This cycle is repeated again and again and thus it flows to the atomizer.

The quantity of fuel delivered is controlled by the rack and pinion arrangement. When the rack moves, the pinion is rotated. The plunger is rotated inside the barrel. It alters the effective stroke of the plunger. Thus, the amount of fuel supplied is varied.

2.2.2 Injection Pump Governor

Depending on the construction these governors are of two types:

i. Gravity controlled centrifugal governors.

ii. Spring controlled centrifugal governors.

There are three commonly used gravity controlled centrifugal governors:

i. Watt governor.

ii. Porter governor.

iii. Pro-ell governor.

The most commonly used spring controlled centrifugal governors is :

i. Hartnell governor.

Watt Governor

It is the primitive governor as used by Watt on some of his early steam engines. It is need for a very slow speed engine and this is why it has now become obsolete. Shown in figure, two arms are hinged at the top of the spindle and two revolving balls new fitted on the other ends the arms.

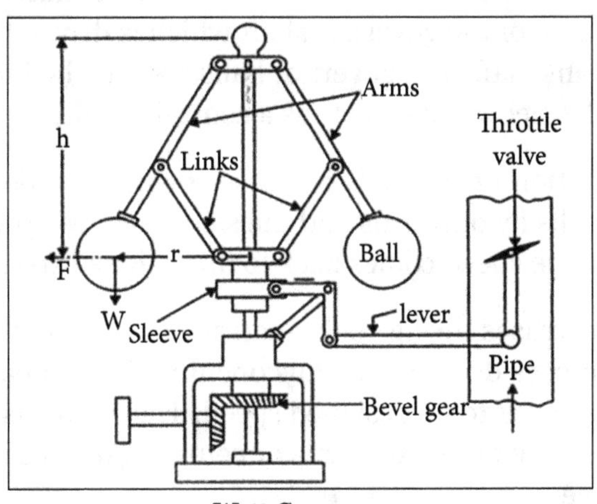

Watt Governor.

One end of each of the links are hinged with the arms, while the other ends are hinged with the sleeve, which may slide over the spindle. The speed of the crankshaft is transmitted to the spindle through a pair of bevel gears by means of a suitable arrangement. So the rotation of the spindle of the governor causes the weights to move away from the centre due to the centrifugal force.

This makes the sleeve to move in the upward direction. This movement of the sleeve is transmitted by the lever to the throttle valve which partially closes or opens the steam pipe and reduces or increases the supply of steam to the engine. So the engine speed may be adjusted to a normal limit.

This governor was used by James Watt in his steam engine. The spindle is driven by the output shaft of the prime mover. The balls are mounted at the junction of the two arms. The upper arms are connected to the spindle and lower arms are connected to the sleeve as shown in Figure:

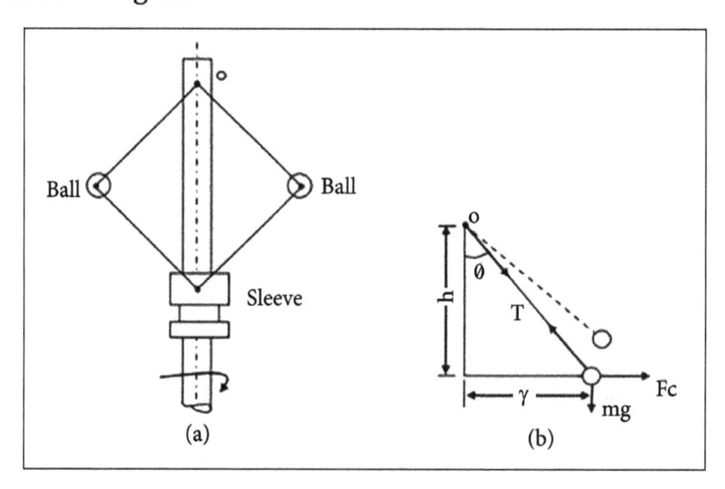

Porter Governor

Figure below shows a Porter governor where two or more masses called the governor balls rotate about the area of the governor shaft which is driven through suitable gearing from the engine crankshaft. The governor balls are attached to the arms. The lower arms are attached to the sleeve which acts as a central weight.

If the speed of the rotation of the balls increases owing to a decrease of load on the engine, the governor balls fly outwards and the sleeve moves upwards thus closing the fuel passage till the engine speed comes back to its designed speed.

If the engine speed decreases owing to an increase of load, the governor balls fly inwards and the sleeve moves downwards thus opening the fuel passage more for oil till the engine speed comes back to its designed speed. The engine is said to be running at its designed speed when the outward inertia or centrifugal force is just balanced by the inward controlling force.

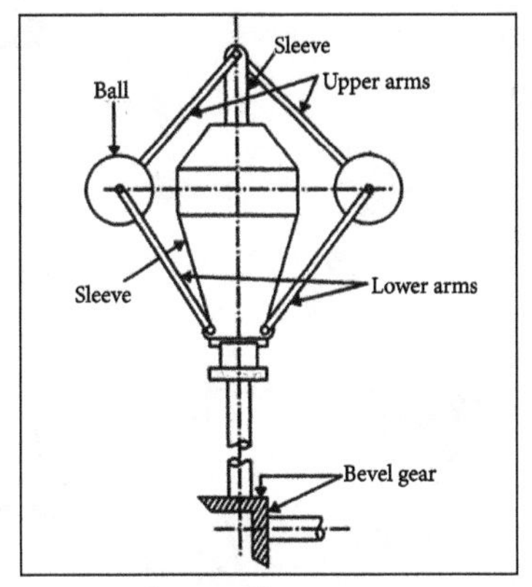

Porter Governor.

A schematic diagram of the porter governor is shown in figure. There are two sets of arms. The top arms OA and OB connect balls to the hinge O. The hinge may be on the spindle or slightly away. The lower arms support dead weight and connect balls also. All of them rotate with the spindle. We can consider one-half of governor for equilibrium.

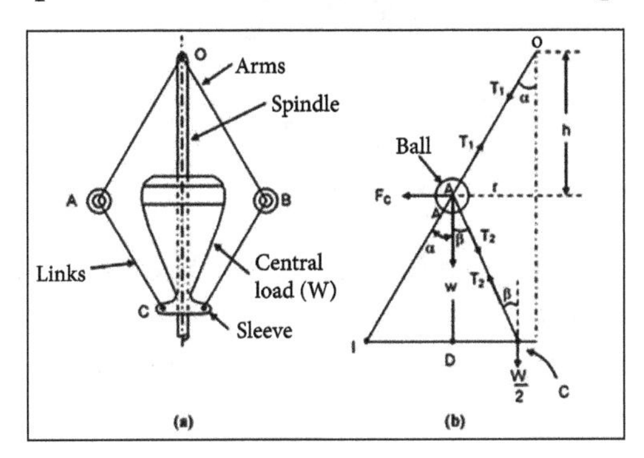

Expression for height in the case of a Watt governor:

The expression of governor height (h) in terms of governor speed (N) is given by,

$$h = \frac{g}{\omega^2} \qquad \because \omega = \frac{2\pi N}{60}$$

$$= \frac{9.81}{\left(\frac{2\pi N}{60}\right)^2} \qquad g = 9.81\ m/s^2$$

$$h = \frac{895}{N^2}$$

Pro-ell Governor

Refer the following figure. It is a modification of porter governor. The governor balls are carried on an extension of the lower arms. For given values of weight of the ball, weight of the sleeve and height of the governor, a Pro-ell governor run at a lower speed than a Porter governor. In order To achieve the equilibrium speed a ball of smaller mass may be used in pro-ell governor.

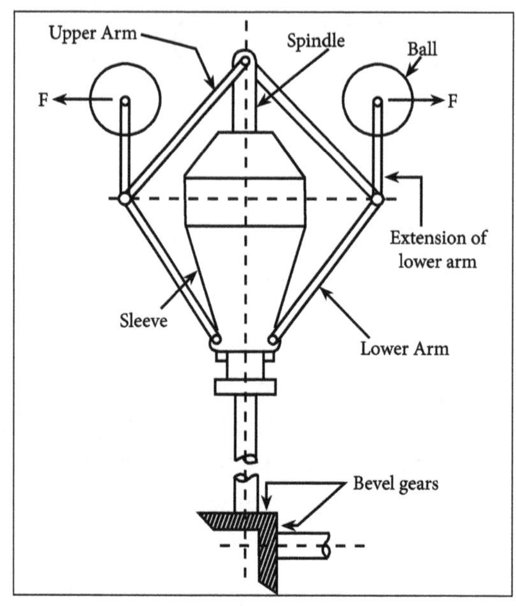

Pro-ell governor.

Hartnell Governor

The hartnell governor is shown in following figure. The two bell crank levers have been provided which can have rotating motion about fulcrums O and O'. One end of each bell crank lever carries a ball and a roller at the end of another arm. The rollers make contact with the sleeve. The frame is connected to the spindle.

A helical spring is mounted around the spindle between frame and sleeve. With the rotation of the spindle, all these parts rotate.With the raising of speed, the radius of rotation of the balls increases and the rollers lift the sleeve against the spring force. With the lessening in speed, the sleeve moves downwards. The movement of the sleeve are transferred to the throttle of the engine through linkages.

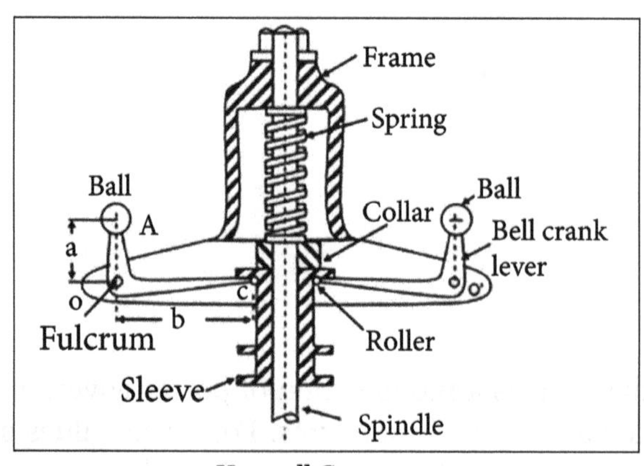

Hartnell Governor.

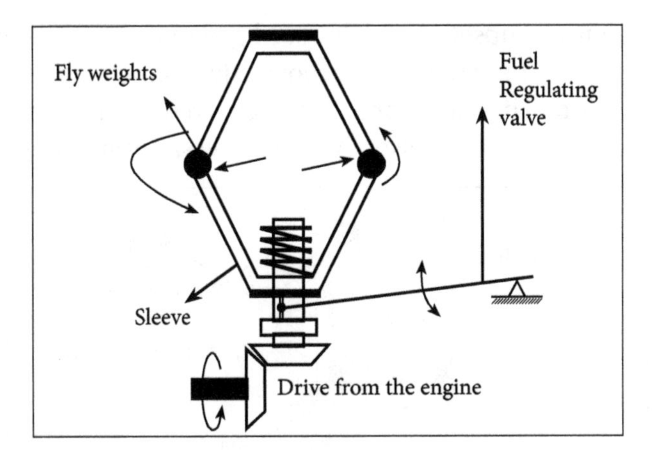

The method of maintaining the turbine speed constant irrespective of the load is known as governing of turbines. The device used for governing of turbines is called Governor.

2.2.3 Mechanical Governor

Mechanical Governors In most of governors are installed on diesel engines used by the Navy, the centrifugal force of rotating weights and the tensions of a helical coil spring are used in governor operation.

In mechanical centrifugal flyweight governors two forces oppose each other. One of these forces is tension spring which may be varied either by an adjusting device or by movement of the manual throttle. The engine produces the other force. The weights, attached to the governor drive shaft, are rotated and a centrifugal force is created when the engine drives the shaft. The centrifugal force varies with the speed of the engine.

Transmitted to the fuel system through a connecting linkage, the tension of the spring tends to increase the amount of fuel delivered to the cylinders. On other hand, the centrifugal force of the rotating weights, through connecting linkage, tends to reduce the quantity of fuel injected. When the two opposing forces are equal or balanced, the speed of the engine remains constant.

Working of Governor

To show how the governor works when the load increases and decreases, let us assume we are driving a truck in hilly terrain. When a truck approaches a hill at a steady engine speed, the vehicle is moving from a set state of balance in the governor assembly with a fixed throttle setting to an unstable condition.

As the vehicle starts to move up the hill at a constant speed, the increased load demands result in a reduction in engine speed. This upsets the state of balance that had existed in the governor. The reduced rotational speed at the engine results in a reduction in speed and therefore, the centrifugal force of the governor weights.

When the state of balance is upset, the high-speed governor spring is allowed to expand, giving up some of its stored energy, which moves the connecting fuel linkage to an increased delivery position. This additional fuel delivered to the combustion chambers would result in an increase in horse power, but not necessarily an increase in engine speed.

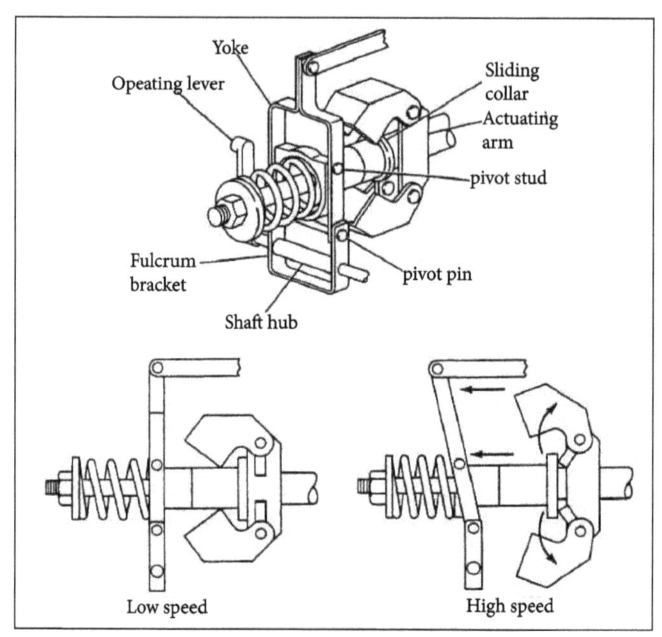

Mechanical (centrifugal) governor.

When the truck moves into a downhill situation, the operator is forced to back off the throttle to reduce the speed of the vehicle, otherwise, the brakes or engine or transmission retarder has to be applied. The operator can also downshift the transmission to obtain additional braking power. However, when the operator does not reduce the throttle position or brake the vehicle mass in some way, an increase in road speed results.

This is due to the reduction in engine load because of the additional reduction in vehicle resistance achieved through the mass weight of the vehicle and its load pushing the truck downhill. This action causes the governor weights to increase in speed and they attempt to compress the high-speed spring, thereby reducing the fuel delivery to the engine.

Engine over speed can result if the road wheels of the vehicle are allowed to rotate fast enough that they in effect, become the driving member. The governor assembly would continue to reduce fuel supply to the engine due to increased speed of the engine.

If over speed does occur, the valves can end up floating and striking the piston crown. Therefore, it is necessary in a downhill run for the operator to ensure that the engine speed does not exceed maximum governed rpm by application of the vehicle, engine or transmission forces.

2.2.4 Pneumatic Governor

1. Throttle Governing.
2. Nozzle Control Governing.
3. By-Pass Governing.

1. Throttle Governing

The steam pressure at inlet to a steam turbine is reduced by throttle process to maintain the speed of the turbine constant at part load and hence this method of governing is called throttle governing. A consider an instant when the load on the turbine increases. As a result the speed of the turbine decreases. The fly balls of the governor will come down.

The fly balls bring down the sleeve. The downward movement of the sleeve will raise the control valve rod.

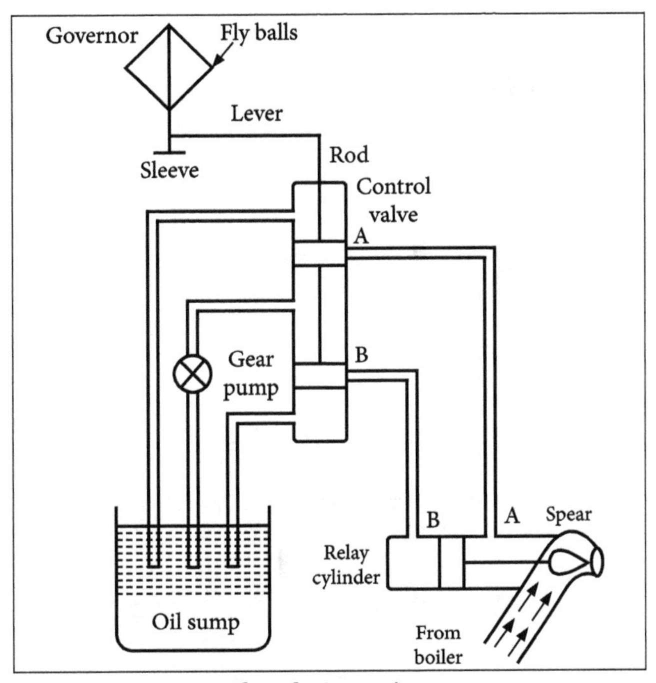

Throttle Governing.

The mouth of the pipe AA will open. Now the oil under pressure will rush from the control valve to the right side of the piston in the relay cylinder through the pipe AA. This will move the piston and spear towards the left which will open more area of nozzle. As a result the steam flow rate into the turbine increases, which in turn brings the speed of the turbine to the normal range.

2. Nozzle Control Governing

Nozzle control governing is used in large power steam turbines to which very high pressure steam is supplied. In this method, the total number of nozzles of a turbine is grouped in a number of groups varying from two to twelve groups and each group of nozzle. Steam supply is, controlled by valves These values are not opened or closed by automatic devices.

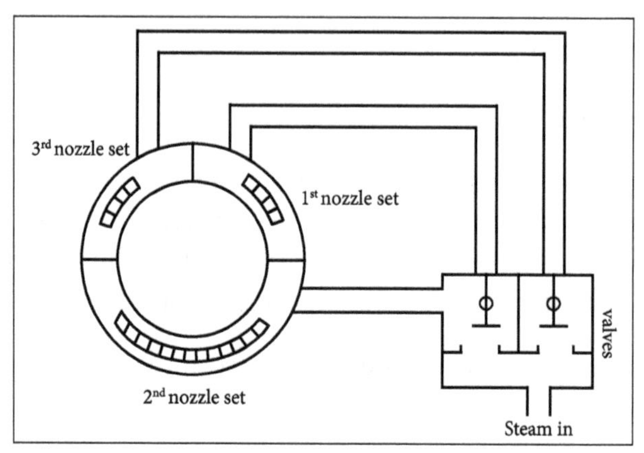

Nozzle Control Governing.

Number of groups of nozzle in operation at a particular instant depends upon the load on the steam turbine.

3. By-Pass Governing

This method is adopted in modern high pressure steam turbines which contain a large number of stages of small mean diameter in high pressure stages. Such turbines are usually designed for a definite load known as economic load at which the efficiency is maximum.

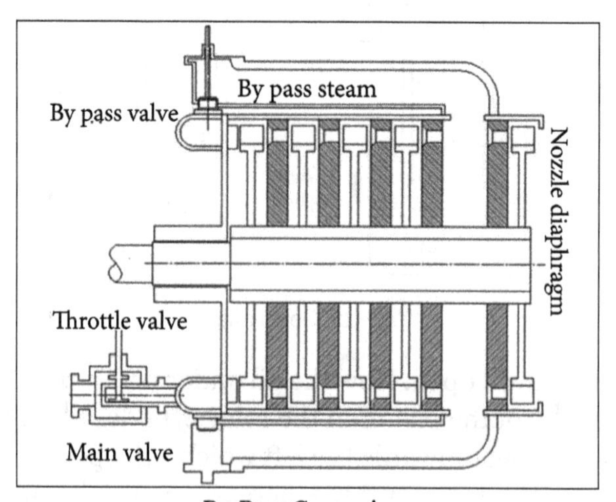

By-Pass Governing.

This load is taken as about 80% of the maximum continuous rating. According to the principle of by-pass governing, some extra quantity of steam by passes to the far down stages of the turbine when the load exceeds the economic load.

2.3 Fuel Injector and Nozzle

Fuel Injector or Atomizer

A metered quantity of fuel from the fuel pump is supplied to the atomizer under high pressure of about 140 bar.

Functions:

1. To inject and atomize the fuel to the required degree.
2. To distribute the fuel that ensures proper mixing.
3. To prevent the injection on the cylinder walls (or) piston top surface.
4. It must start and stop instantaneously.

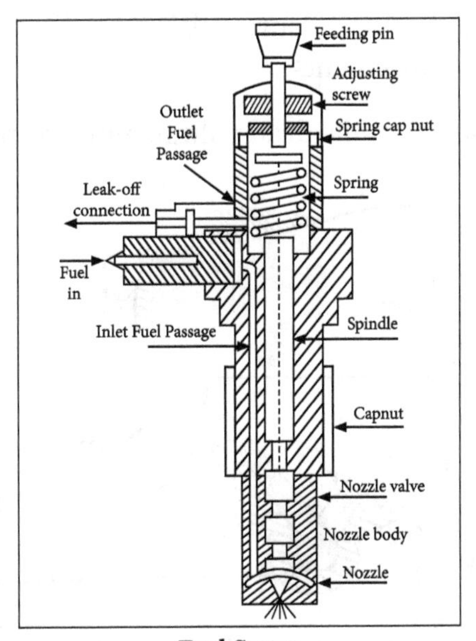

Fuel Spray.

The fuel injector in which a spring loaded nozzle valve opens only at high pressure at about 135 bar to 170 bar.

The spring loaded valve is lifted up due to high pressure fuel entering at the inlet passage and the fuel is injected into the cylinder through the nozzle in the form of fine spray. As the fuel is injected, the fuel pressure falls and hence the nozzle valve rests on its seat under the spring force. Thus the valve is automatically closed and the fuel supply is stopped.

Any backage or overflow of fuel at the plunger of nozzle valve occurs the fuel is sent back to the fuel tank through the outlet passage. The tension of the controlling spring can be adjusted by a adjusting screw.

Nozzle

Design of the diesel fuel injector nozzle is critical to the performance and emissions of modern diesel engines. Some of the important injector nozzle design parameters include details of the injector seat, the injector sac and nozzle hole size and shape. These features not only affect the combustion characteristics of the diesel engine, they can also affect the stability of the emissions and performance over the lifetime of the engine and the mechanical durability of the injector.

All nozzles must spray fuel such that they meet the requirements of the performance and emissions goals of the market for which the engine is produced regardless of details of the fuel system design:

1. Common rail nozzle operates under more demanding tribological conditions and must be better designed to prevent leakage.

2. Unit injector/unit pump pressure pulsing conditions create more demanding fatigue strength requirements.

3. Pump-line-nozzle hydraulic dead volume must be minimized.

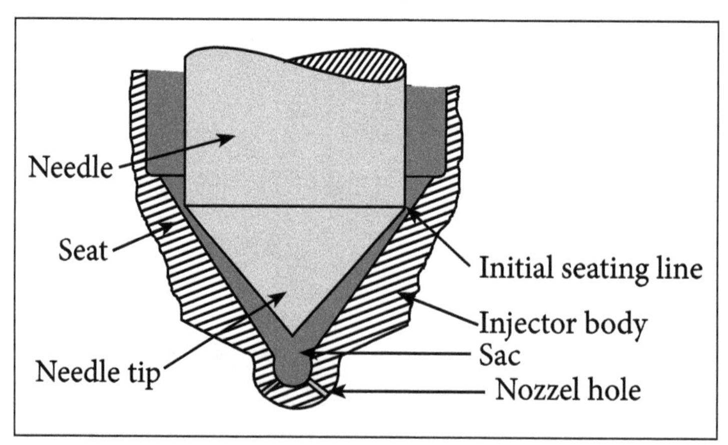

Basic Diesel Injector Nozzle with Single Cone Seat.

Induction of Fuel in SI Engine

1. The task of the engine induction and fuel systems is to prepare from ambient air and fuel in the tank an air-fuel mixture that satisfies the requirement of the engine.

2. This preparation is to be carried out over entire engine operating regime.

3. In principle, the optimum air-fuel ratio for an engine is that which give the required power output with the lowest fuel consumption.

4. It should also ensure smooth and reliable operation.

Time Factor

1. Time factor calculated based on the energy balance between the surrounding air and the liquid droplet and the assumption that the heat transferred is a fraction of the available energy.

2. The size of droplet and its energy will decide the rate of evaporation.

Droplet Size Distribution

1. The droplet size distribution in sprays is the crucial parameter needed for the fundamental analysis of the transport of mass, momentum and heat in evaporation.

2. Parameter determines the quality of the spray and consequently influences to a significant extent the processes of emissions in combustion.

3. Detailed experimental data is used to develop distribution functions.

4. To obtain the detailed quantitative information of the sprays, a two-component Phase Doppler Anemometry (PDA) is used.

5. This performs the simultaneous measurements of the droplet velocity and size and the volume flux.

2.4 Injection in SI Engine

Induction

The task of the engine induction and fuel systems is to prepare from ambient air and fuel in the tank an air-fuel mixture that satisfies the requirement of the engine. This preparation is to be carried out over entire engine operating regime.

In principle, the optimum air-fuel ratio for an engine is that which give the required power output with the lowest fuel consumption. It should also ensure smooth and reliable operation.

The fuel Induction systems for SI engine are classified as:

i. Throttle body Fuel Injection Systems.

ii. Multi Point Fuel Injection Systems.

Merits of Fuel Injection in the SI Engine

1. Absence of Venturi – No Restriction in Air Flow/Higher Vol. Eff./Torque/Power.

2. Hot Spots for Preheating cold air eliminated/Denser air enters.
3. Manifold Branch Pipes Not concerned with Mixture Preparation (MPI).
4. Better Acceleration Response (MPI).
5. Fuel Atomization Generally Improved.
6. Use of Greater Valve Overlap.
7. Use of Sensors to Monitor Operating Parameters/Gives Accurate Matching of Air/fuel Requirements: Improves Power, Reduces fuel consumption and Emissions.
8. Precise in Metering Fuel in Ports.
9. Precise Fuel Distribution Between Cylinders (MPI).
10. Fuel Transportation in Manifold not required (MPI) so no Wall Wetting.
11. Fuel Surge During Fast Cornering or Heavy Braking Eliminated.
12. Adaptable and Suitable For Supercharging (SPI and MPI).
13. Increased power and torque.

2.4.1 Electronic Injection Systems

The figure shows the fuel injected system L. Jetronic with air flow metering developed by Robert Based Corp. It consist of the following units:

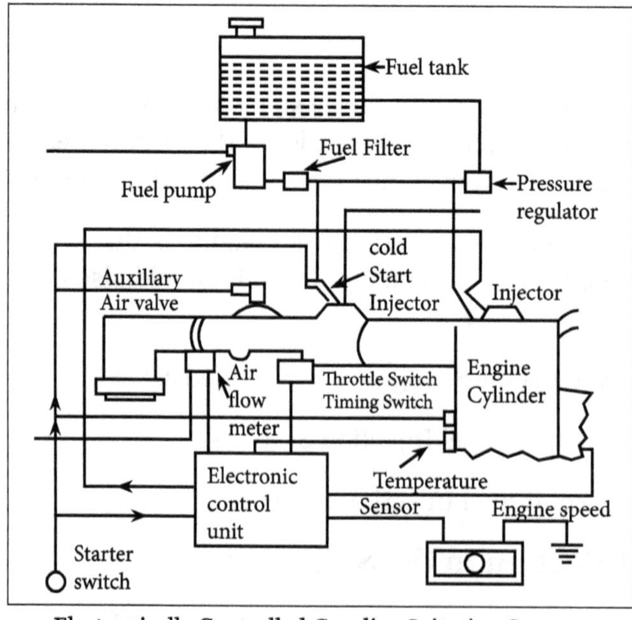

Electronically Controlled Gasoline Injection System.

1. Fuel Delivery System.
2. An Induction System.
3. Sensors and Air Flow Control System.
4. Electronic Control Unit.

1. Fuel Delivery System

1. It consists of an electrically driven fuel pump which draws fuel from a fuel tank. The pump forces the oil through a filter into a line at the end of which is situated a pressure regulator, which in turn is connected to intake manifold.
2. The pressure regulator keeps the pressure difference between the fuel pressure and the manifold pressure constant, so that the quantity of fuel injected is dependent in the injector open time only.

2. Air Induction System

1. After passing the air filter, the incoming air flows through an air flow methane, which generates a voltage signal (depending on the quantity of air flow).
2. Just behind the throttle valve is fitted a cold start magnetic injection valve, which injects additional fuel for cold start. This valve also supplies the extra fuel needed during warm up period.
3. An auxiliary valve by-pass the throttle valve and the throttle valve supplies the extra air required for idling In addition to rich-air fuel mixture. This extra air increases the engine speed after cold start to acceptable idling speed.
4. To the throttle valve is attached a throttle switch equipped with a set of contacts which generate a sequence of voltage signals during the opening of throttle valve, the voltage signals results in injection of additional fuel required for acceleration.

3. Electronic Control Unit

The sensors are incorporated to measure the operating data at different locations. The data measured by the sensors are transmitted to the electronic control unit which computes the amount of fuel injected during each engine cycle. The amount of fuel injected is varied by varying the injector opening time only.

The sensors used are dependent upon:

1. Manifold pressure.
2. Engine speed.
3. Temperature at the intake manifold.

4. Injection Time

1. For every revolution of the Camshaft, the fuel is injected twice, each injection contributing half of a fuel quantity required for engine cycle.
2. The injectors at different phases of the operating cycle are operating simultaneously.

Merits and Demerits of Electronic Fuel Injection

Merits

1. It uses engine sensors, computers and solenoid operated fuel injectors to meter and inject the right amount of fuel into the engine cylinder.
2. Electronic devices monitor and control engine operation.
3. Sensors controls the engine temperature, air flow, air inlet, pressure, camshaft position and knock correctly.
4. Because of direct spray discharge into each inlet port, acceleration response is better.
5. Multi-point injection does not require time for fuel transportation in the intake manifold and there is no manifold wall melting.
6. It is possible to use greater inlet and exhaust valve overlap without poor idling, loss of fuel or increased exhaust pollution.

Demerits

1. Initial cost of equipment is high.
2. More electrical and Mechanical components may go wrong.
3. More Bulky and heavy.
4. Due to pumping and metering of the fuel there is increased mechanical and hydraulic noise.
5. Injection equipment may be elaborately complicated to handle and impossible to service.

2.4.2 Multi-Point Fuel Injection (MPFI) System

Multi-point fuel injection systems are the most common and usually have an injector per cylinder located in each individual manifold runner.This configuration gives much better control of fueling and better emissions since the fuel can be metered more closely and there is less opportunity for the fuel spray to condense or drop out of the airflow since it is introduced as small streams rather than one large one.

If Closer to the inlet valve the fuel injection takes place, then better the economy and transient throttle. Most systems use one injector per cylinder but on certain engines there are only two inlet ports since two cylinders share a siamesed port. In this case multi-point would mean two injectors, one per inlet port.

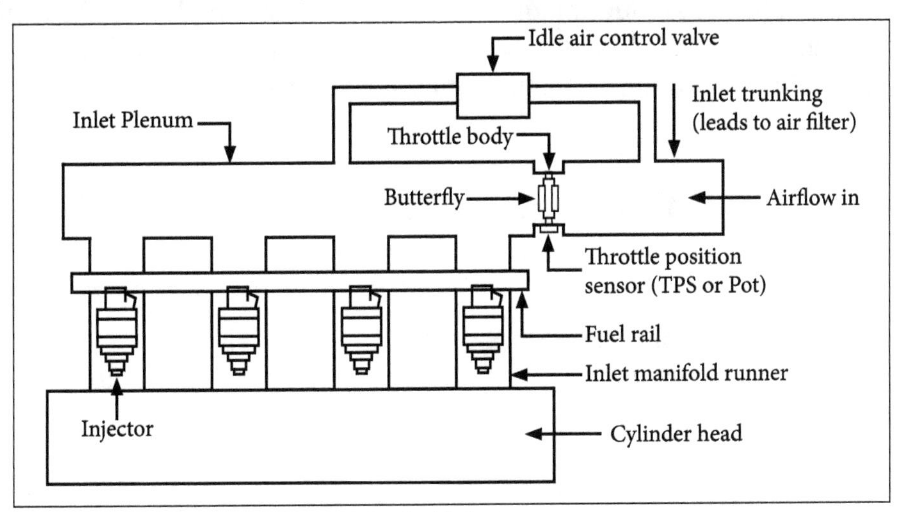

Multi-point fuel injection systems.

With multi-point or multi injector systems there is scope for timing the injection of fuel to better suit the engine's duty cycle. If the EMS knows the relative position of each cylinder within the engines cycle then it can fire the injectors at the optimum time for that cylinder. This is known as sequential injection. Sometimes the EMS will only have knowledge of the crank position rather than the duty cycle position and in this case it can optimise for a pair of cylinders.

This is known as semi-sequential or grouped injection. Some EMS systems ignore the crank and cycle position when injecting fuel. They fire all of the injectors at the same time once per revolution and this is known as batched injection. There is no penalty to pay power-wise when using batched injection. However, grouped and sequential injection give a slight edge on economy and transient throttle and emissions.

Advantages of Multi Point Fuel Injection System

1. More uniform air-fuel mixture will be supplied to each cylinder, hence the difference in power developed in both cylinder is minimum.

2. The mileage of the vehicle is improved.

3. No need to crank the engine twice or thrice in case of cold starting happening in the carburetor system.

4. Immediate response, in case of sudden acceleration and deceleration.

5. The vibrations produced in MPFI engines is very less, due to this life of the engine component is increased.

6. More accurate amount of air-fuel mixture will be supplied in these injection system. As a result complete combustion will take place. This leads to effective utilization of fuel supplied and hence low emission level.

2.5 Functional Divisions of MPFI System and Injection Timing

Mono point and Multi point Injection Systems: In petrol injection system, the fuel is injected into the intake manifold through fuel injection valves. There are two basic arrangements i.e., multi point and mono point injection.

1. Multi Point Injection System

It is also called as port injection system. In this system, there is an injection valve for each engine cylinder as shown in figure (a). Each injection valve is placed in the intake port near the intake valve as shown in figure (b). The main advantage of this system is that it allows more time for the mixing of air and petrol.

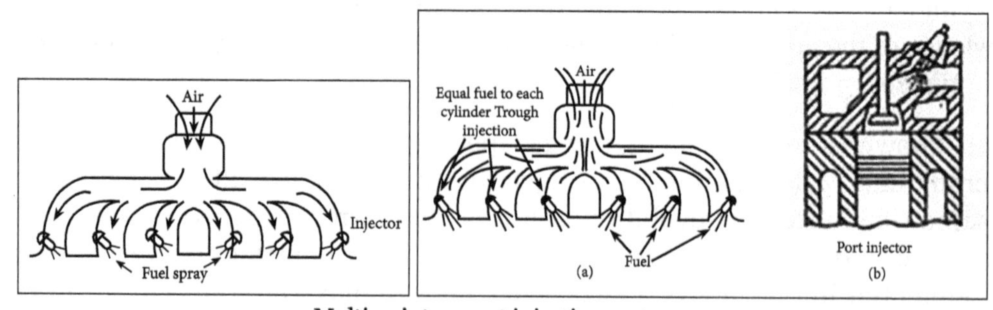

Multi point or port injection system.

2. Mono Point Injection or Throttle Body Injection System

This system is also called as Throttle Body Injection. In this system, an injection valve is positioned slightly above each throat of the throttle body as shown in figure. The injection valve sprays fuel into the air just before it passes through the throttle valve and enters the intake manifold.

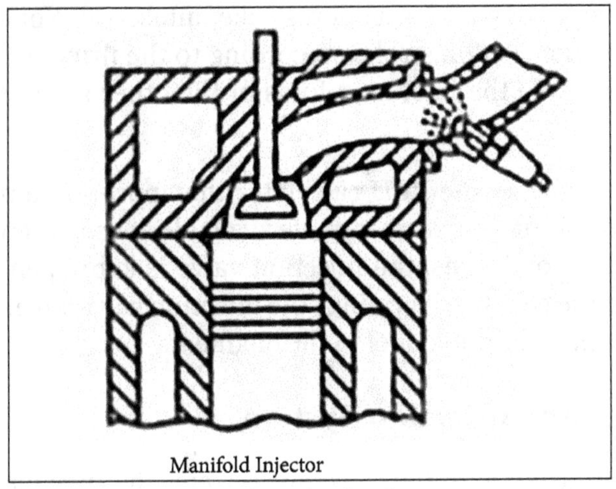

Mono point Injection System.

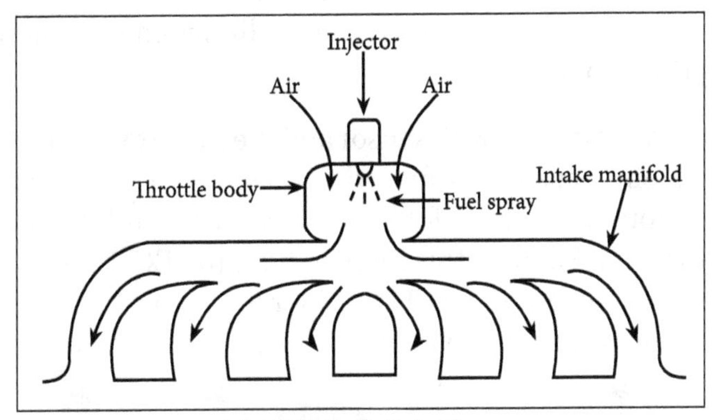

Throttle Body Injection.

This method simplifies the construction of the engine block. Also, it does not obstruct hot spots near the valves affecting cooling water jacket size at that place. Moreover, it requires only one circuit in the computer to control injection which simplifies the construction of electronic control unit. Thus, it reduces the cost of the system.

Injection Timing

Consider a cylinder of a four cylinder engine. The fuel is injected into the inlet manifold of each cylinder at different timings. The timing at which the injection of the fuel takes place inside the inlet manifold is called injection tinting. The injection timing for one cylinder of this four cylinder engine is described below.

In one cylinder, the piston moves up from BDC (Bottom Dead Centre) to TDC (Top Dead Centre) during the exhaust stroke. just before the piston reaches TDC during this exhaust stroke, injection of the fuel takes place into the inlet manifold of this cylinder at about 60° crank angle before TDC.

This injected fuel mixes with the air in the air intake chamber. Thus the air-fuel mixture is obtained. At the beginning of the suction stroke, intake valve opens and the air-fuel mixture is sucked by the suction stroke. According to the firing order, the injection of the fuel takes place inside the inlet manifolds of the other three cylinders at various timings.

In this four cylinder engine, the ECU calculates the appropriate injection timing for each cylinder and the air fuel-mixture is made available at each suction stroke. In order to meet the operating conditions, the injection valve is kept open for longer time by ECU. An example, if the car is accelerating , the injection valve will be opened for longer time, in order to supply additional fuel to the engine.

2.5.1 Group Gasoline Injection System

In an engine having group gasoline injection system, the injectors are not activated individually, but are activated in groups. In a four-cylinder engine also there are two groups, each group having two injectors. Ina six-cylinder engine, there are two groups, each group having three injectors.

A figure (a) shows a block diagram with sensor and the Electronic Control Unit (ECU), for a group injection system. Sensors for detecting pressure in the manifold, engine speed in rpm, throttle position, intake manifold air temperature and the coolant temperature send information to the ECU. With this information, the ECU computes the amount of gasoline that the engine needs. The ECU then sends signals to the injectors and other parts of the system. The timing of the injectors is decided by the engine-speed sensor.

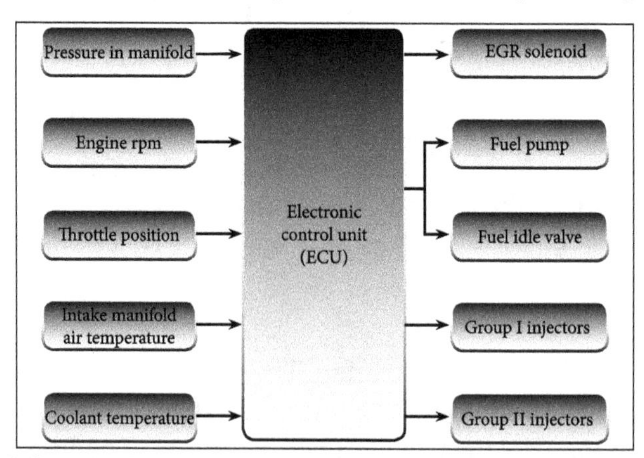

(a) Block diagram showing the sensors and ECU for group injection system.

The injectors are divided into two groups. Based on the signals from the speed sensor, the ECU activates one group of injectors. Subsequently, The ECU activates the other group of injectors. For example, the injector grouping for a six-cylinder engine is shown in figure (b). Injectors for cylinders 1, 3 and 5 open at the same time and inject gasoline into the intake manifold. After these injectors clone, the injectors for the cylinders 2, 4 and 6 open and inject gasoline.

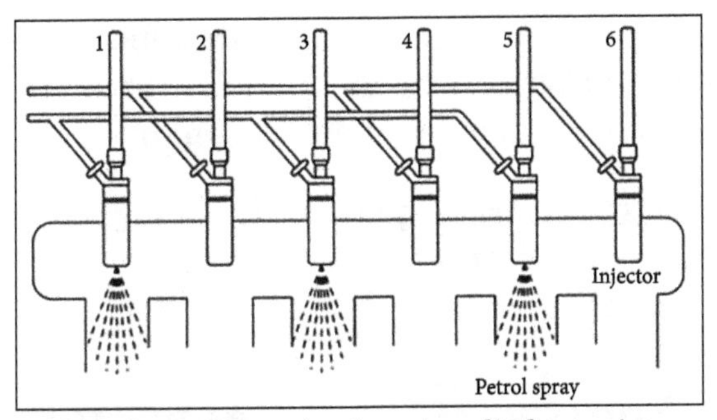

(b) Injectors grouping in a six-cylinder engine.

Figure(c) shows a port injection using the electronic group fuel-injection system for an eight-cylinder engine. Eight injectors are connected to a fuel system and are divided into two groups, each group having four injectors. Each group of injectors is alternately turned on by the ECU.

When the crankshaft makes two revolutions. Thus it is seen that the modern engines are controlled more and more by electronics and the days are not far off when electronics may completely take over leaving bare mini-mum room for any mechanical or manual control.

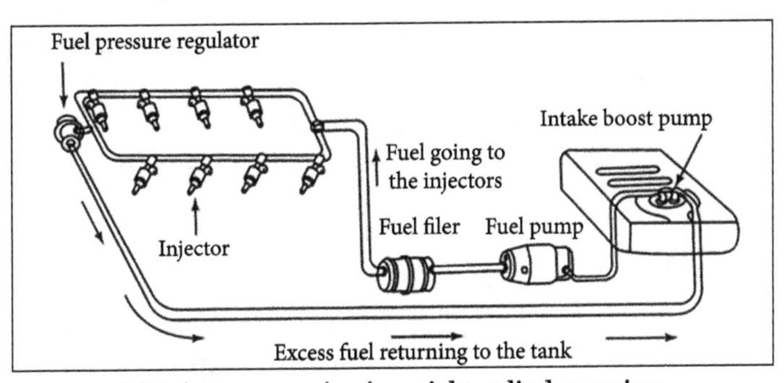

(c) Injector grouping in a eight-cylinder engine.

2.5.2 Electronic Diesel Injection System

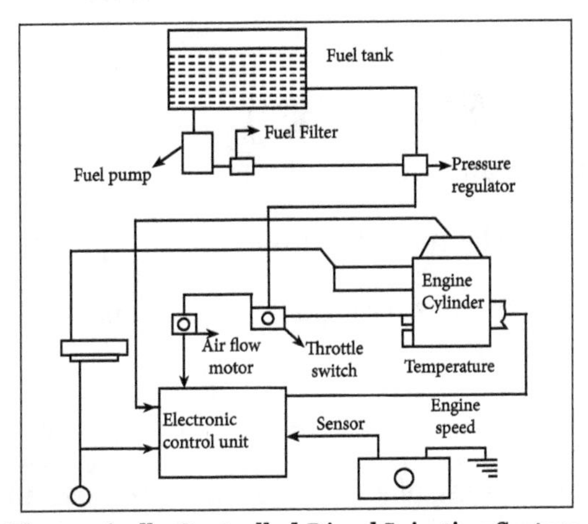

Electronically Controlled Diesel Injection System.

It consists of electrically driven fuel pump which draw fuel from a fuel tank.The pressure regulation keeps the pressure and the manifold pressure constant so that the quantity of fuel injected is dependent on injector open time only.

An auxiliary valve provides the extra air required for idling. This extra air increases the engine speed after cold start to acceptable idling speed.

2.6 Ignition: Energy Requirement for Ignition and Requirements of an Ignition System

Energy Requirement for Ignition

The total enthalpy required to cause the flame to be self sustaining and promote ignition, is given by the product of the surface area of the spherical flame and the enthalpy per unit area. It is reasonable to assume that the basic requirement of the ignition system is that it should supply this energy within small volume.

Further, ignition should occur in a time interval sufficiently short to ensure that only a negligible amount of energy is lost to establish the flame. In view of this last mentioned condition, it is apparent that the rate of supply of energy is as important as the total energy supplied. A small electric spark of short duration would appear to meet most of the requirements for ignition.

A spark can be caused by applying a sufficiently high voltage between two electrodes separated by a gap and there is a critical voltage below which no sparking occurs. This critical voltage is a function of the dimension of the gap between the electrodes, the fuel-air ratio and the pressure of the gas. Additionally, the manner in which the voltage is raised to the critical value and the configuration and the condition of the electrodes are important in respect of the energy required. An ignition process obeys the law of conservation of energy.

Hence, it can be treated as a balance of energy:

i. That is provided by an external source.

ii. That is released by chemical reaction.

iii. That is dissipated to the surroundings by means of thermal conduction, convection and radiation.

Requirements of an Ignition system

A smooth and reliable functioning of an ignition system is essential for reliable working of an engine. The requirements of such an ignition system are:

i. It should provide a good spark between the electrodes of the plugs at the correct timing.

ii. It should function efficiently over the entire range of engine speed.

iii. It should be light, effective and reliable in service.

iv. It should be compact and easy to maintain.

v. It should be cheap and convenient to handle.

vi. The interference from the high voltage source should not affect the functioning of the radio and television Deceivers inside an automobile.

2.6.1 Conventional Ignition Systems and Modern Ignition Systems (TCI and CDI)

Conventional Ignition Systems

The jump-spark system used today in the internal combustion engine has been gradually developed through the stages of hot wire, break spark trembler coil, with each step showing a definite improvement over its predecessor. The two jump-spark ignition generator systems in use today are the battery-coil and the magneto, the latter is confined mainly to the small engines used on motor cycles and lawn mowers.

Coil-Ignition System

In 1908 the battery-inductive ignition system was introduced by C.F. Kettering of Delco, but only in the mid 1920s it could achieve its commercial status as a successor to the magneto. Up to that time very few vehicles used a battery, hence the magneto was common being a self-contained ignition generator. With the introduction of electric lighting, use of a battery becomes necessary. Because of this as well as the difficulty in starting of the magneto-ignited engine, the battery inductive system commonly known as coil ignition was introduced.

Conventional Coil-ignition Circuits

The circuit shown in Figure is for a coil-ignition system with main components. The heart of the system is the ignition coil, which transforms the low-tension (LT), 12 V supply given by the battery to the high-tension (HT) with voltage needed to produce a spark at the spark plug.

The coil has primary and secondary windings forming two complete circuits that make up the complete system. Primary, the LT circuit, is supplied by the battery and secondary, the HT circuit incorporates the distributor and spark plugs.

The end of the secondary winding in the coil is earthed, which is achieved by connecting the winding either to the LT coil terminal (normally the negative) or to an additional coil terminal that is linked by an external cable to earth. The latter arrangement of coil is called an insulated return (IR) coil to distinguish it from the common earth return (ER) type and is needed on a vehicle using IR system. The contact breaker interrupts the primary DC current to induce the HT voltage into the secondary winding at the instant the spark is required.

To obtain the precise timing of the spark the break in the primary circuit is required to be sudden and to avoid this critical stage , a capacitor is installed 'across' the contact breaker. The operation of a coil-ignition system is based on the principles of mutual induction and the transformer action. When both the ignition switch and contact breaker are closed, a current of about 3 A flows through the primary winding of the coil creating a strong magnetic flux around the winding.

The contact breaker is opened at the appropriate time by a cam driven at the speed of the engine camshaft, i.e. half crankshaft speed. This breakage of the primary circuit causes a sudden collapse of the magnetic flux in the coil and induces an emf into the secondary winding, which has about 60 times as many turns as the primary winding.

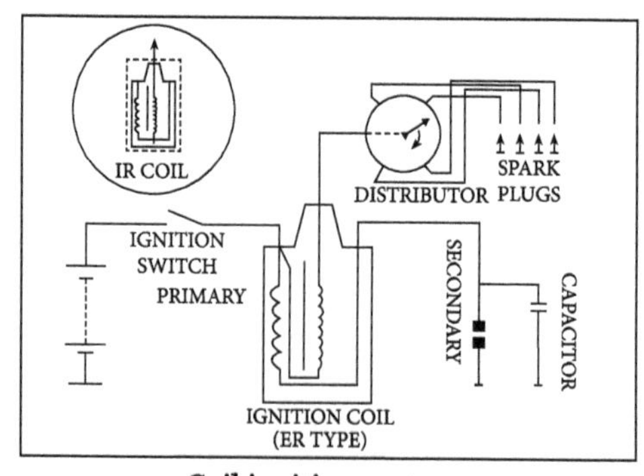

Coil ignition system.

The transformer action, combined with the effect of the self-induced voltage in the primary, steps-up the voltage to that required to produce a spark at the plug. However, with the increase in the secondary voltage there is a proportional decrease in the current. The secondary winding is connected to the negative LT coil terminal, due to which the primary and secondary windings are arranged in series. This connection is called the auto-transformer connection, which adds the self-induced emf in the primary to the mutually induced emf in the secondary providing a higher output.

In a single-cylinder engine, a highly insulated lead is used to convey the HT current directly to the spark plug. But, a distributor is necessary in a multi-cylinder engine to allocate the HT current to the appropriate spark plug. The distributor is a HT rotary switch comprising of a distributor and a rotor arm, rotating at camshaft speed. The plug leads are connected to brass electrodes in the cap maintaining the firing order of the cylinders.

A lead from the coil tower makes contact with a carbon brush that rubs on a brass blade forming part of the rotor arm. An automatic advance mechanism, installed adjacent to the contactor breaker, alters the timing of the spark to suit the engine speed and load. It alters the spark timing by moving both the cam and the base plate on which the contact breaker is mounted. The unit, called ignition distributor incorporates the distributor, contact breaker and automatic advance mechanism.

Modern Ignition Systems CDI And TCI

CDI

The capacitive discharge ignition (CDI) system is a solid state ignition system. It is standard equipment in many applications and has improved the reliability of modern small gasoline engines. The only moving parts in a CDI system are the permanent magnets in the flywheel. Refer to figure (a) to progressively trace current flow through the various electronic components M a typical CDI system.

As the flywheel magnets rotate across the CDI module laminations, they induce a low voltage alternating current (ac) in the charge coil. The ac passes through a rectifier and is changed to direct current (dc), which travels to the capacitor, where it is stored. The CDI ignition module is compact and maintenance. The only moving parts in a CDI system are the flywheel magnets.

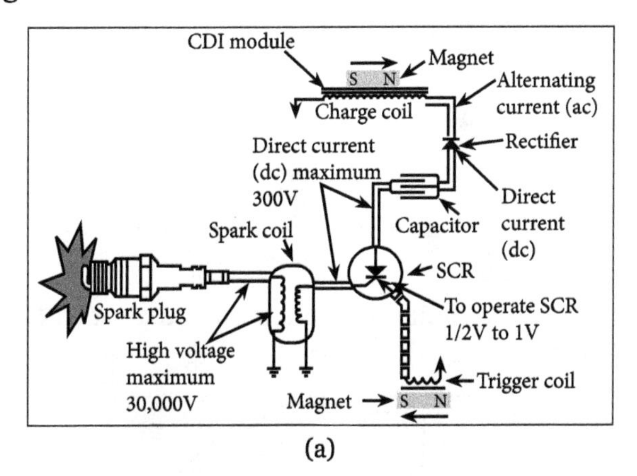

(a)

The flywheel magnets induce a low-voltage alternating current in the charge coil. As the alternating current passes through the rectifier, it is changed to direct current. The direct current continues to the capacitor, where it builds up a charge. When the capacitor nears its full charge, the flywheel magnets induce a small current in the trigger coil. The current briefly activates the silicon controlled rectifier (SCR), which allows the 300V stored in the capacitor to discharge through the primary windings of the spark coil. This induces a much higher voltage in the secondary windings, which fires the spark plug.

When the silicon controlled rectifier is triggered, the 300V dc stored in the capacitor travels to the spark coil. At the coil, the voltage is stepped up instantly to a maximum of 30,000V. This high voltage current is discharged across the spark plug gap.

In figure (b), the flywheel magnets rotate approximately 351° before passing the CDI module laminations and induce a small electrical charge in the trigger coil. At starting speeds, this electrical charge is just great enough to turn on the silicon controlled rectifier (SCR) in a retarded firing position (9° BTDC). This provides for easy starting.

In figure (c), when the engine reaches approximately 800 rpm, advanced firing begins. The flywheel magnets travel approximately 331°, at which time enough voltage is induced in the trig-ger coil to energize the silicon controlled rectifier in the advanced firing position (29° BTDC).

Operation of Transistor-Controlled Ignition (TCI) System

The individual components that make up the transistor-controlled ignition (TCI) system are:

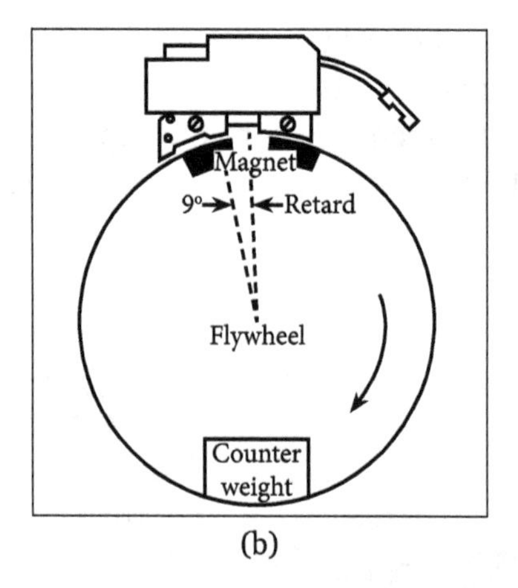

(b)

At low speeds, the flywheel magnets induce a same current in the trigger coil, which turns on a silicon rectifier at V BTDC for easy starting.

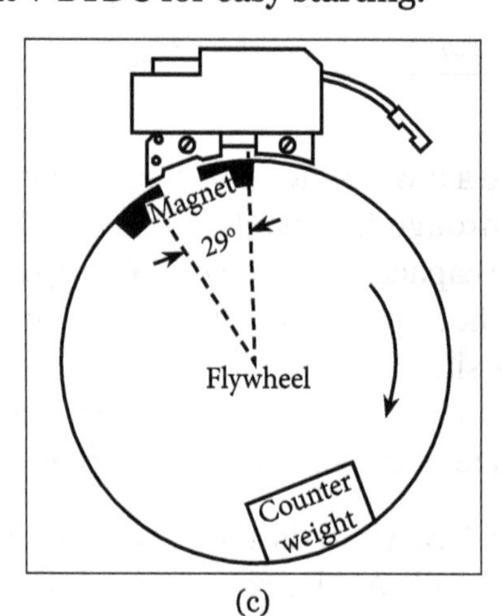

(c)

At 800 rpm, stronger trigger coil current turns on the silicon rectifier at 29° UDC for satisfactory ignition during normal engine operation. a chart Is given in figure (d) . Study the function of each part carefully. There are a variety of transistor-controlled circuits. Each has its own unique characteristics and modifications.

Figure (e) illustrates a typical circuit for a transistor-controlled ignition. As the engine flywheel rotates, the magnets on the flywheel pass by the ignition coil. The magnetic field around the magnets induces current in the primary windings of the ignition coil.

The base circuit of the ignition system has current flow from the coil primary windings, common grounds, resistor (R1), base of the transistor (T1), emitter of the transistor (T1) and back to the primary windings of the ignition coil. Current flow for the collector circuit in figure (e) is from the primary windings of the coil, common grounds, collector of transistor (T1), emitter of transistor (T1) and back to the primary windings. When the flywheel rotates further, the induced current in the coil primary increases.

Diode (D1, D2)	A —▶|— K	Allows one way current from Anode "A" to Cathode "K" as rectifier.
Flywheel		Provides magnetic flux to primary windings of ignition coil.
High-tension lead		Conducts high voltage current in secondary windings to spark plug.
Ignition coil		Generates primary current, and transforms primary low voltage to secondary high voltage
Ignition switch		No spark across gap of spark plug when switch is at "STOP" position.
Resistor (R1, R2)		Resists current flow.
Spark plug		Ignites fuel-air mixture in cylinder.
Thyristor (S)	A —▶|— K, G	Switches from blocking state to conducting state when trigger current/voltage is on gate "G".
Transistor	B, C, E	Very small current In the base circuit (B to E) controls and amplifies very large current in the collector circuit (C to E). When the base current is cut, the collector current is also cut completely.

(d)

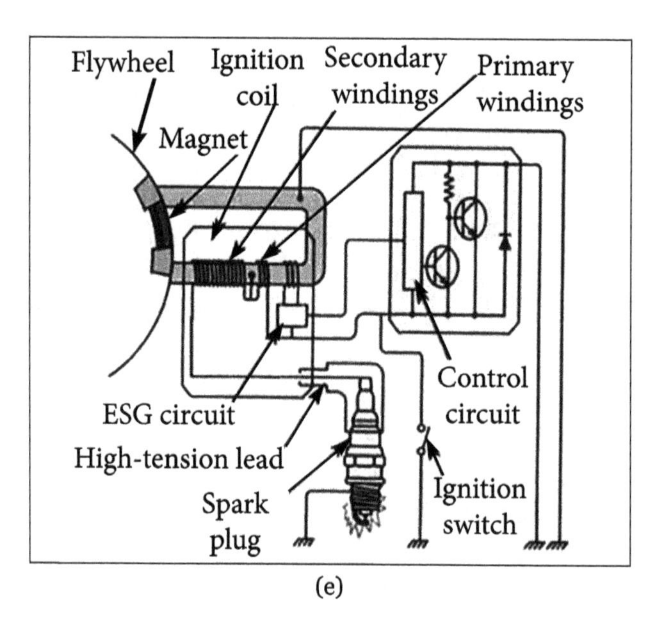

(e)

When the current is high enough, the control circuit turns on, this causes transistor (T2) to turn on and conduct. A strong magnetic field forms around the primary winding of the ignition coil. The trigger circuit for this ignition system consists of the primary windings, common grounds, control circuit, base of transistor (T2) and emitter of transistor (T2). When transistor (T2) begins to conduct current, the base current flow is cut. This causes the collector circuit to shut off and the transistor (T1) stops conducting current.

When transistor (T1) stops conducting, current stops flowing through the primary of the ignition coil. This causes the primary magnetic field to collapse across the secondary windings of the ignition coil. High voltage is then induced into the secondary winding to fire the spark plug. The secondary circuit includes the coil secondary windings, spark plug wire, spark plug and common grounds returning to the coil secondary.

When the ignition switch is off, the primary circuit is grounded to prevent the plug from firing. Diode (Dl) is installed in the circuit to protect the TCI module from damage. The ESG circuit shown in Figure (e) is used to retard the ignition timing. At high engine rpm, the ESG circuit conducts. This bypasses the trigger circuit and delays when the current reaches the base of transistor (T2).

2.6.2 Firing Order, Ignition Timing and Spark Advance Mechanism

Firing Order of Cylinders

Cylinder firing order improves the distribution of the fresh charge in the manifold to the cylinders and helps the release of the exhaust gases, while at the same time suppresses torsional vibrations. These conditions are as follows:

i. Successive cylinders firing allows a recovery of charge in the manifold and minimizes interference between adjacent or nearby cylinders. Normally cylinders from opposite end of the manifold are chosen or from alternate cylinder banks in V engines to draw alternately. This arrangement, however, becomes difficult as the number of cylinders decreases.

ii. Separating successive cylinders, which are exhausting, are even more important than for induction. It is because if the exhaust periods overlap with the cylinders, exhaust-gas back pressure may prevent escaping of products of combustion from the cylinders.

iii. Power impulses cause winding up of the crankshaft. In addition, if the natural torsional oscillations of the shaft coincide with these disturbing impulse frequencies, torsional vibrations may take place. Therefore, in general, it is desirable to have successive power impulses to alternate ends of the crankshaft.

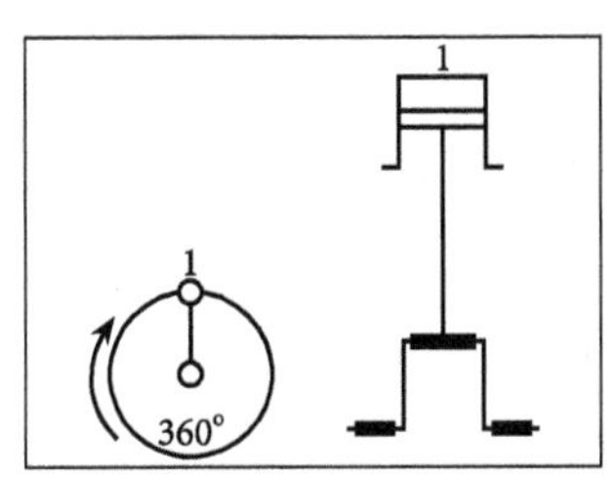

Single-cylinder arrangement.

The Spark Energy and Duration (Ignition timing)

With a homogeneous mixture in the cylinder, spark energy of the order of 1 m.1 and a duration of a few micro-seconds would suffice to initiate the combustion process. However, in practice, circumstances are less than the ideal. The pressure, temperature and density of the mixture between the spark plug electrodes have a considerable influence on the voltage required to produce a spark.

The spark energy and duration are to be of sufficient order to initiate combustion under the most un favorable conditions expected in the vicinity of the spark plug over the complete range of engine operation. Usually, if the spark energy exceeds 40 ml and the duration is longer than 0.5 ms, reliable ignition is obtained. If the resistance of the deposits on the spark plug electrodes are sufficiently high, the loss of electrical energy through these deposits may prevent the spark discharge.

Spark Advance Mechanism

It is obvious from the above discussion that the point in the cycle where the spark occurs must be regulated to ensure maximum power and economy at different speeds and loads and this must be done automatically. The purpose of the spark advance mechanism is to assure that under every condition of engine operation, ignition takes place at the most favorable instant of time, i.e., most favorable from a standpoint of engine power, fuel economy and minimum exhaust dilution.

By means of these mechanisms the advance angle is accurately net so that ignition occurs before the top dead center point of the piston. The engine speed and the engine load are the control quantities required for the automatic adjustment of the ignition timing.

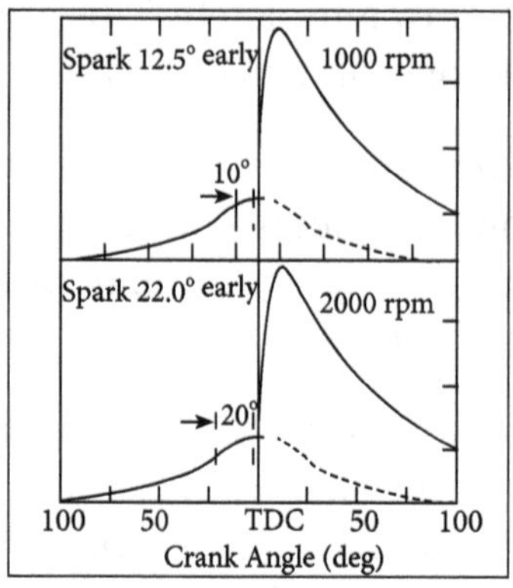

Effect of spark advanced on pressure-crank angle diagram.

Most of the engines are fitted with mechanisms which are integral with the distributor and automatically regulate the optimum spark advance to account for change of speed and load. The two mechanisms used are:

i. Centrifugal advance mechanism.

ii. Vacuum advance mechanism.

2.7 Combustion: Stages of Combustion in SI and CI Engines

The relatively rapid chemical combination of hydrogen and carbon in fuel with oxygen resulting in liberation of energy in the form of heat is known as combustion.

The following conditions are necessary for combustion to take place:

1. The presence of combustible mixture.
2. Some means to initiate mixture.
3. Stabilization and propagation of flame in Combustion Chamber.

Stages of Combustion

There are three stages of combustion in SI Engine:

1. Ignition lag stage.
2. Flame propagation stage.
3. After burning stage.

Ignition Lag

There is a certain time interval between instant of spark and instant where there is a noticeable rise in pressure due to combustion. The time interval in the process of chemical reaction during which molecules get heated up to self ignition temperature, get ignited and produce a self propagating nucleus of flame is called Ignition lag.

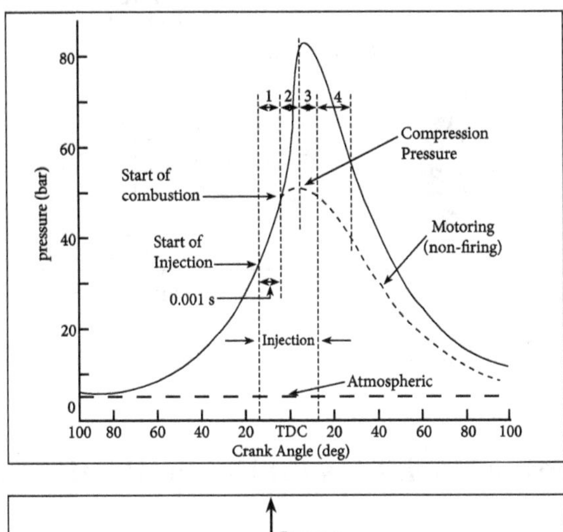

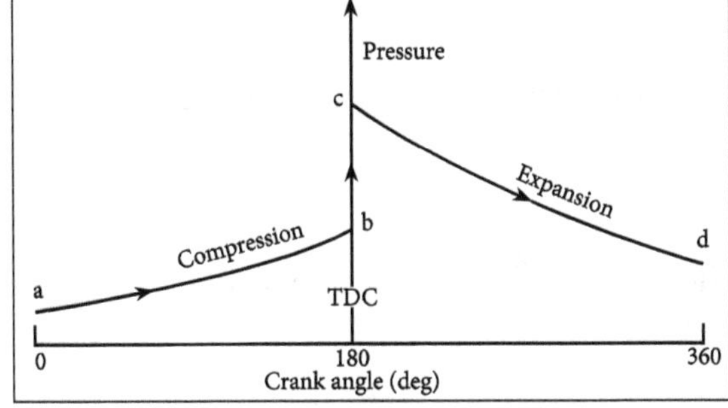

Flame Propagation

Once the flame is formed at "b", it should be self sustained and must be able to propagate through the mixture. This is possible when the rate of heat generation by burning is greater than heat lost by flame to surrounding. After the point "b", the flame propagation is abnormally low at the beginning as heat loss is more than heat generated. Therefore pressure rise is slow as mass of mixture burnt is small.

Therefore it is necessary to provide angle of advance 30 to 35° If the peak pressure to be attained 5-10° after TDC. The time required for crank to rotate through an angle $\theta 2$ is known as combustion period during which propagation of flame takes place.

After Burning

Combustion will not stop at point "c" but continue after attaining peak pressure at point "c" which is known as after burning. This generally happens when the rich mixture is supplied to engine.

Abnormal Combustion: When the combustion gets deviated from the normal behavior resulting in loss of performance or damage to the engine.

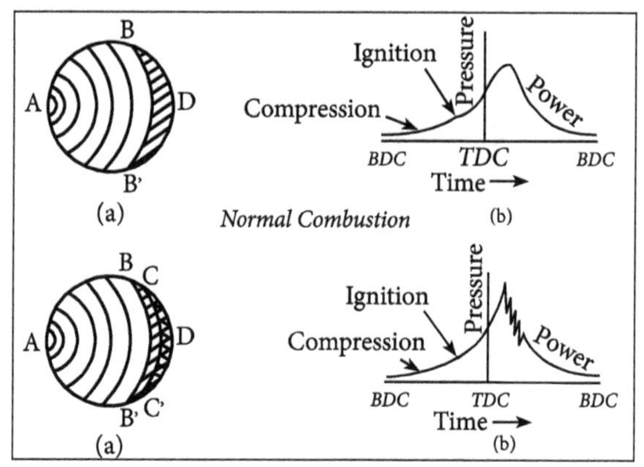

Combustion with detonation normal and abormal combustion.

Normal Combustion: When the flame travels evenly or uniformly across the combustion chamber.

2.7.1 Effects of Engine Variables on Flame Propagation and Ignition Delay

Combustion is dependent upon the rate of propagation of flame front (or flame speed).

Flame Front

Boundary or front surface of the flame that separates the burnt charges from the unburnt one is known as flame front.

Flame Speed

The speed at which the flame front travels is called flame speed. Flame speed affects the combustion phenomena, pressure developed and power produced. The burning rate of mixture depends on the flame speed and shape/contour of combustion chamber.

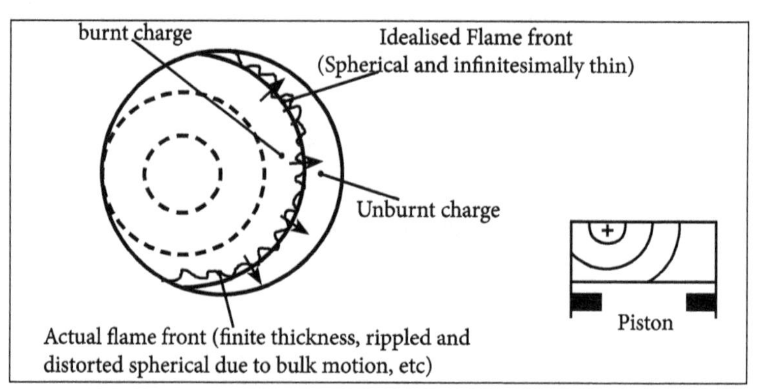

Factors Affecting Flame Speed (FS)

Turbulence

It helps in mixing air and fuel and accelerates chemical reaction. A lean mixture can be burnt without difficulty.

Engine Speed

When engine speed increases, flame speed increases due to turbulence, swirl, squish and tumble.

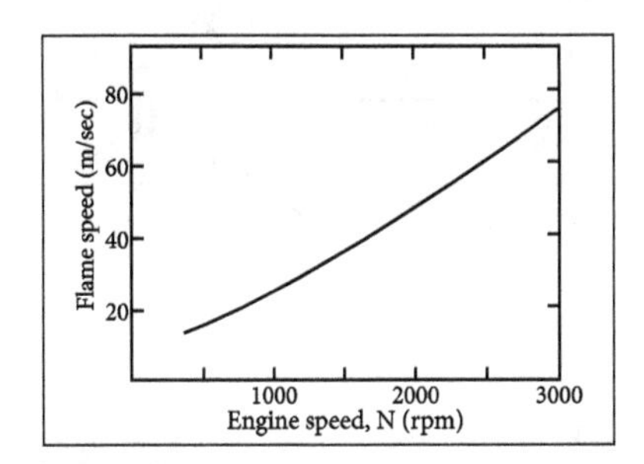

Compression Ratio (CR)

A higher CR increases the pressure and temperature of mixture. This reduces the initial phase of combustion and hence less ignition advance is needed. High p and T of the compressed mixture speed up the 2nd phase of combustion. Increased CR reduces the clearance volume and hence the density of charge. This further increases the peak pressure and temperature, reducing the total combustion duration. Thus, an engine with higher CR have higher flame speeds.

Inlet Temperature and Pressure

FS increases with increase of inlet temperature and pressure. A higher values of inlet temperature and pressure form a better homogeneous mixture, which helps in increasing the FS.

Fuel-Air Ratio

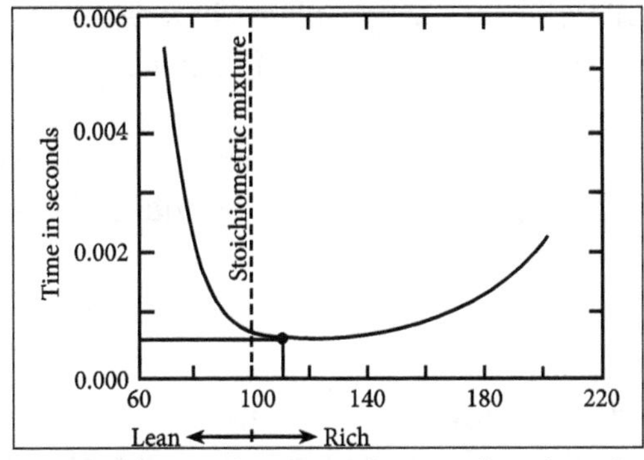

Average flame speed in CC of an SI engine as a function of air-fuel ratio.

The highest flame speeds are obtained with slightly rich mixture (point A). When the mixture is leaner or richer, flame speed decreases.

Effect of mixture strength on burning rate Is given by,

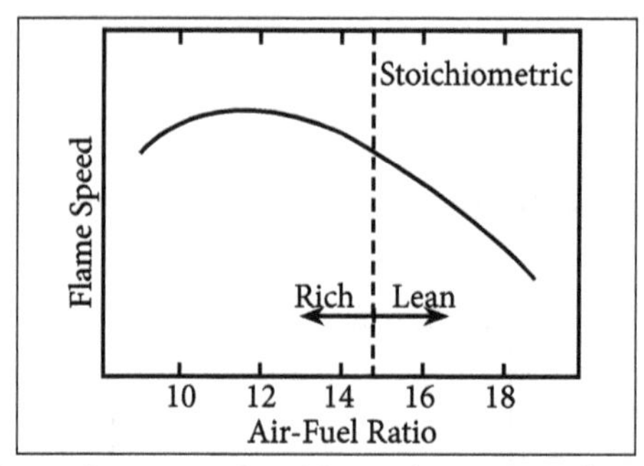

Lean air-fuel mixtures have slower flame speeds with maximum speed occurring when slightly rich at an equivalence ratio near 1.2.

2.8 Abnormal Combustion, Pre-Ignition and Detonation

Abnormal Combustion

Due to excessively weak mixtures combustion may be slow or may be mis-timed. These are however obvious. There are two combustion abnormalities, which are less obvious;

1. The first of these is pre or post ignition of the mixture by incandescent carbon particles in the chamber. This will have the effect of reducing the work transfer.

2. The second abnormality is generally known as knock and is a complex condition with many facets. A simple explanation shows that knock occurs when the unburnt portion of the gas in the combustion chamber is heated by combustion and radiation so that its temperature becomes greater than the self ignition temperature.

If normal progressive combustion is not completed before the end of the induction period then a simultaneous explosion of the unburnt will occur. This explosion is accompanied by a detonation (pressure) wave which will be repeatedly reflected from the cylinder setting up a high frequency resonance which gives an audible noise. The detonation wave causes excessive stress and destroys the thermal boundary layer at the cylinder causing overheating.

Pre-Ignition

i. Pre-ignition is the ignition of the homogeneous mixture in the cylinder, before the timed ignition spark occurs, followed by the local overheating of the combustible mixture. For premature ignition of any local hot-spot to occur in advance of the timed spark on the combustion stroke it must attain a minimum temperature of something like 700- 800°C.

ii. Pre-ignition is initiated by some overheated Projecting part such as the sparking plug electrode, exhaust valve head, metal corners in the combustion chamber, carbon deposit or protruding cylinder head gasket rim etc.

iii. However, pre-ignition is also caused by persistent detonating pressure shock waves scoring away the stagnant gases which normally protect the combustion chamber walls. The resulting increased heat flow through the walls, raises the surface temperature of any protruding poorly cooled part of the chamber and this there-fore provides focal point for pre-ignition.

iv. The initiation of ignition and the propagation of the flame front from the heated hot-spot is similar to that produced by the spark-plug when it fires, the only difference between the hot-spot and spark plug is their respective instant of ignition. Thus, the sparking plugs provides a timed and controlled moment of ignition whereas the heated surface forming the hot-spot builds up to the ignition temperature during each compression stroke and therefore the actual instant of ignition is unpredicable.

The early ignition created by pre-ignition extends the total time and the burnt gases remain in the cylinder and therefore increases the heat transfer on the chamber walls, as a result, the self-ignition temperature will occur earlier and on both successive compression stroke.

Consequently, the peak cylinder pressure (which normally occurs at its optimum position of 10°-15° after T.D.C.) will progressively advance its position to T.D.C. where the cylinder pressure and temperature will be maximized.

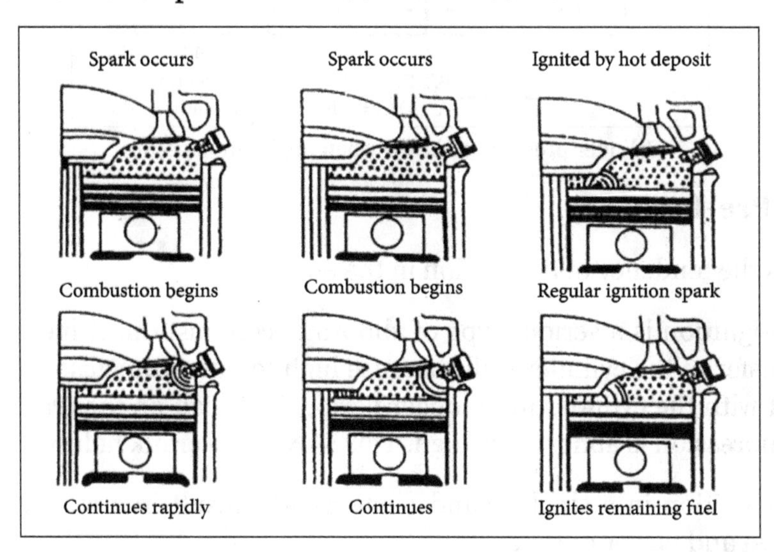

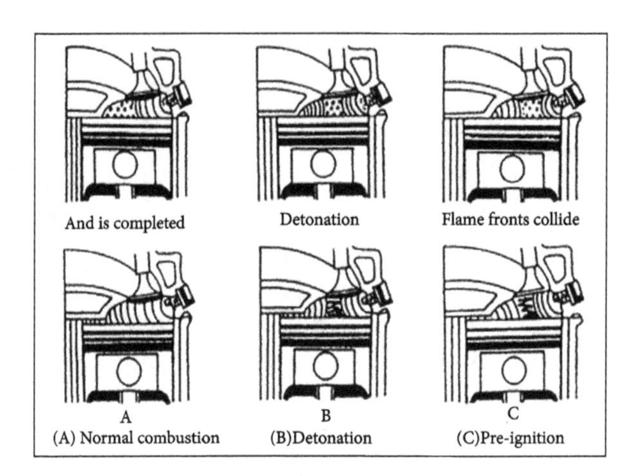

(A) Normal combustion (B)Detonation (C)Pre-ignition

The accumulated effects of en extended combustion time and rising peak cylinder pressure and temperature cause the self-ignition temperature to creep further and further ahead of T.D.C. and with it, peak cylinder pressure, which will now take place before T.D.C. so that negative work will be done in compressing the combustion products.

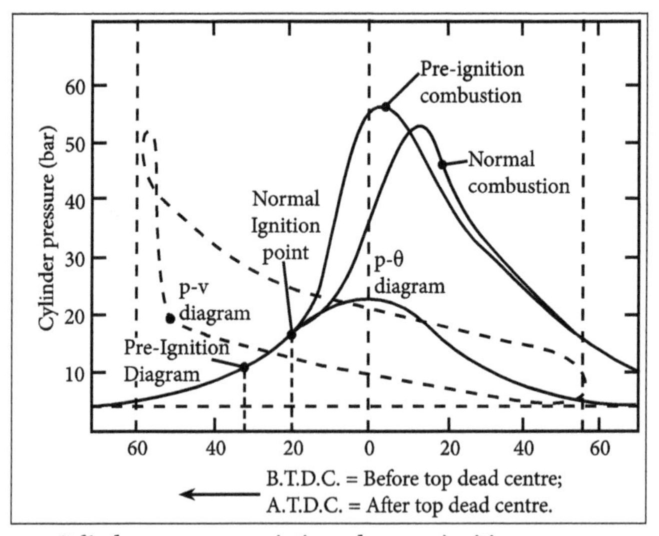

Cylinder pressure variation when pre ignition occurs.

Effects of Pre-Ignition

1. It is the tendency of detonation in the engines.
2. Pre-ignition is a serious type of abnormal combustion. It increases the heat transfer to the cylinder walls because high temperature gases remain in contact with the cylinder for a longer period. The load on the crankshaft during compression is abnormally high. This may cause crank failure.
3. Pre-ignition in a single-cylinder engine will result in a steady reduction in speed and power output.

4. The real undesirable effects of pre-ignition are when it occurs only in one or more cylinders in a multi-cylinder engine. Under these conditions, when the engine is driven hard, the unaffected cylinders will continue to develop their full power and speed and then will drag the other piston or pistons, which are experiencing pre-ignition and are producing negative work, to and fro until eventually the increased heat generated makes the pre-igniting cylinders' pistons and rings sizes.

Thus, the danger of the majority of cylinders operating efficiently while one or more cylinders are subjected to exceptive pre-ignition is that the driver will only be aware of a loss in speed and power and therefore try to work the engine harder to compensate for this, which only intensifies the pre-ignition situation until seisure occurs.

The following points are worth noting:

1. Pre-ignition is not responsible for abnormally high cylinder pressure, but there can be a slight pressure rise above the normal due to the ignition point and, therefore, the peak pleasure creeping forward to the T.D.C. position where maximum pressure occurs.

2. If pre-ignition occurs at the same time as the timed sparking plug tree, combustion will appear as normal. Therefore, if ignition is switched-off, the engine would continue to operate at the same speed as if it were controlled by the conventional timed spark, provided the self-ignition temperature continues to occur at the same point.

3. Over heated spark plugs which are the main causes of pre-ignition should be carefully avoided in the engines.

Detonation

At present the amount of power that can be developed in the cylinder of a petrol engine is fixed by the liability of a fuel to detonate, i.e. just before the flame has completed its course across the combustion chamber and remaining unhand charge fires throughout its spontaneously without external assistance.

The result is a tremendously rapid and local increase in pressure which sets up pressure waves that hit the cylinder walls with such violence that the walls emit a sound like a ping. It is the ping that manifests detonation. Thus a very sudden rise of pressure during combustion a companied by metallic hammer like sound is called detonation.

The region in which detonation occurs is farthest removed from the sparking plug and is named the 'detonation zone' and even with severe detonation this zone is rarely more than one quarter the clearance volume.

Process of Detonation or Knocking

The process or phenomenon of detonation or knocking may be explained by referring to the Figure, which shows the cross-section of the combustion chamber with flame advancing from the spark plug location A. The advancing flame front compresses the end charge BB'D farthest from the spark plug, thus raising its temperature.

The temperature of the end charge also increases due to heat transfer from the hot advancing flame front. Also some preflame oxidation may take place and charge leading to further increase in temperature. charge BB'D reaches its auto-ignition temperature and remains for some time to complete the preflame reaction, the charge will auto ignite leading to knocking combustion.

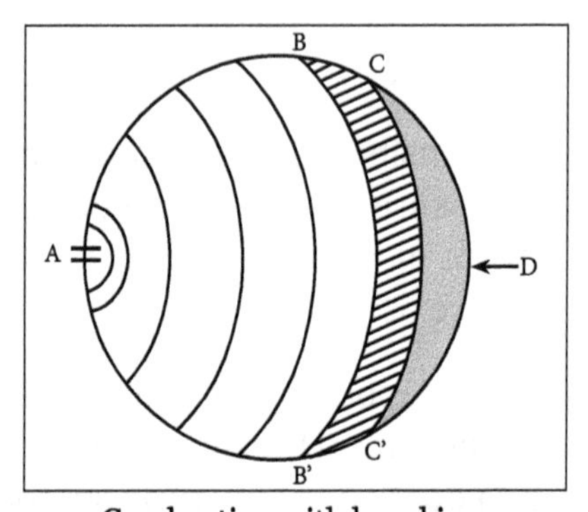

Combustion with knocking.

During the preflame reaction period the flame front could move from BB' to CC' and the knock occurs due to auto-ignition of the charge ahead of CC'. Here we have combustion unaccompanied by flame, producing a very high rate of pressure rise. The pressure-time diagram of detonating combustion in S.I. engines is drawn and labeled below:

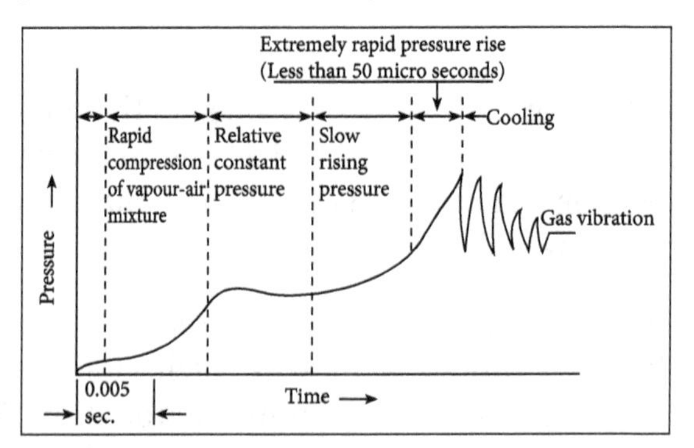

The 'intensity of detonation' will depend mainly upon the amount of energy contained in the 'end-mixture' and the rate of chemical reaction which releases it in the form of heat and a high intensity pressure-wave. Thus, earlier in the combustion process the detonation commences, the more unburnt end-mixture will be available to intensify the detonation. As little as 6 per cent of the total mixture charge when spontaneously ignited will be sufficient to produce a very violent knock.

2.9 Theory of Detonation: Effect of Engine Variables on Detonation and Control of Detonation

There are two general theories of knocking/detonation:

1. The auto-ignition theory.
2. The detonation theory.

1. Auto-Ignition Theory

Auto-ignition refers to initiation of combustion without the necessity of a flame. The auto-ignition theory of knock assumes that the flame velocity is normal before the on-set-of auto-ignition and that gas vibrations are created by a number of end-gas elements auto-igniting almost simultaneously.

2. DetonationTheory

In the auto-ignition theory, it is assumed that the flame velocity is normal before the onset of auto-ignition whereas in detonation theory a true detonating wave formed by preflame reactions has been proposed as the mechanism for explosive auto-ignition. Such a shock wave would travel through the chamber at about twice the sonic velocity and would compress the gases to pressures and temperatures where the reaction should be practically instantaneous.

In fact knocking or detonation is a complex phenomenon and no single explanation may be sufficient to explain it fully.

Effects of Detonation

1. Noise and roughness.
2. Mechanical damage.
3. Carbon deposit.
4. Increase in heat transfer.
5. Decrease in power output and efficiency.
6. Pre-ignition.

Control of Detonation

The detonation can be controlled or even stopped by the following methods:

1. Increasing engine r.p.m.
2. Retarding spark.

3. Reducing pressure in the inlet manifold by throttling.
4. Making the ratio too lean or too rich, preferably latter.
5. Water injection. Water injection increases the delay period as well as reduces the flame temperature.
6. Use of high octane fuel can eliminate detonation. High octane fuels are obtained by adding additives known as dopes.

Factors Affecting Detonation/Knocks

The likelihood of knock is increased by any reduction in the induction period of combustion and any reduction in the progressive explosion flame velocity. Particular factors are listed below:

1. Fuel choice: A low self-ignition temperature promotes knock.
2. Induction pressure: Increase of pressure decreases. The self-ignition temperature and the induction period. Knock will tend to occur at full throttle.
3. Engine speed: Low engine speeds will give low turbulence and low flame velocities (combustion period is constant in angle).c,d knock may occur at low speed.
4. Ignition timing: Advanced ignition timing increases peak pressures and promotes knock.
5. Mixture strength: Optimum mixture strength gives high pressures and promotes knock.
6. Compression ratio: High compression ratios increase the cylinder pressures and promotes knock.
7. Combustion chamber design: Poor design gives long flame paths, poor turbulence and insufficient cooling all of which promote knock.
8. Cylinder cooling: Poor cooling raises the mixture temperature and promotes knock.

2.10 Diesel Knock and Methods to Control Diesel Knock

Diesel knock is the sound produced by the very rapid rate of pressure rise during the early part of the uncontrolled second phase of combustion. The primary cause of an excessively high pressure rise is due to a prolonged delay period. An extensive delay period can be due to the following factor:

i. A low design compression ratio permitting only a marginal self-ignition temperature to be reached.

ii. A low combustion pressure due to worn piston rings or bad seating valves.

iii. Poor fuel ignition quality, that is a low cetane number fuel.

iv. A poorly atomized fuel spray preventing early ignition to be established.

v. An inadequate injector needle spring load producing coarse droplet formation.

vi. A very low air intake temperature in cold weather and during cold starting.

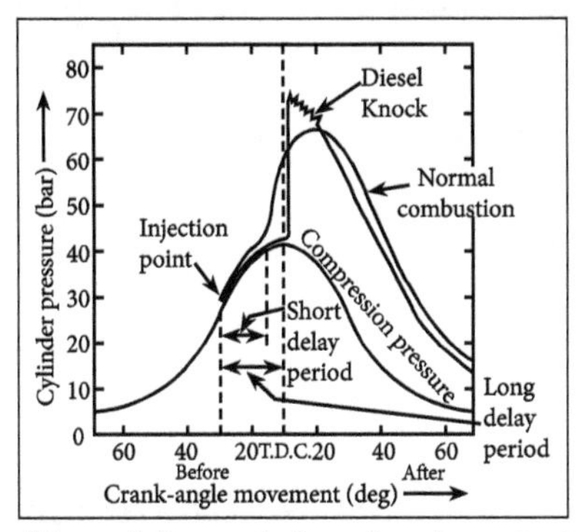

Effect of short and long delay period on the characteristic p-θ diagram.

A very long ignition lag after injection causes a large proportion of the fuel discharge to enter the cylinder and to atomise before ignition and the propagation of burning actually occurs. Accordingly, when combustion does commence, a relative amount of heat energy will be released almost immediately, this correspondingly produces the abnormally high rate of pressure rise, which is mainly responsible for rough and noisy combustion process under these condition.

It has been observed generally, that provided the rate of pressure increase does not exceed 3 bar per degree of crank-angle movement, combustion will be relatively smooth, whereas between a 3 and 4 bar pressure rise there is a tendency to knock and, above this rate of pressure rise, diesel knock will be prominent.

Differences in the knocking Phenomenon of the SI and C.L Engines

The following are the differences in the knocking phenomena of the SI. and C.I. engines:

1. In the SI engine, the detonation occurs near the end of combustion whereas in the CI engine detonation occurs near the beginning of combustion.

2. The detonation in the S.I. engine is of a homogeneous charge causing very high rate of pressure rise and very high maximum pressure. In the Cl. engine, the fuel and air are imperfectly mixed and hence the rate of pressure rise is normally lower than that in the detonating part of the charge in the S.I. engine.

3. In the Cl engine, the fuel is injected into the cylinder only at the end of the compression stroke, there is no question of pre-ignition as in Si. engine.

4. In the SI engine, it is relatively easy to distinguish between knocking and non-knocking operation as the human ear easily finds the distinction.

5. Factors that tend to reduce detonation in the Si. engine increase knocking in the C.I. engine.

Methods of Controlling Diesel knock (Reducing Delay Period)

The diesel knock can be controlled by reducing delay period. The delay is reduced by the following:

1. High charge temperature.
2. High fuel temperature.
3. Good turbulence.
4. A fuel with a short induction period.

2.10.1 Requirements of Combustion Chambers

The three basin requirements of a good S.l. engine combustion chamber are:

1. High power output with minimum octane requirements.
2. High thermal efficiency.
3. Smooth engine operation.

1. High Power Output Requires

1. High compression ratio. The compression ratio is limited by the phenomenon of detonation. It depends upon the design of combustion chamber and fuel quality. Any change in design that improves the anti-knock characteristics of a combustion chamber permits the use of higher compression ratio which should result in higher output and efficiency.

2. Small or no excess air.

3. A complete utilization of the air - no dead pockets.

4. An optimum degree of turbulence. Turbulence is induced by inlet flow configuration or squish. Turbulence induced by squish in preferable to inlet turbulence as the volumetric efficiency is not affected.

5. High volumetric efficiency.

This is achieved by having large diameter, valve timings and straight passage ways by streamlining the combustion chamber so that flow is with lesser pressure drop. This means more charge per stroke and proportionate increase in the power output.

2. High Thermal Efficiency Requires

1. High compression ratio.
2. Compact combustion chamber for small heat loss during combustion, reduced Flame travel and thus less combustion time loss.

3. Smooth Engine Operation Requires

1. Moderate rate of pressure rise during combustion.
2. Absence of detonation, by proper location of spark plug and exhaust valve, by satisfactory cooling of spark plug points (to avoid pro-ignition) and of exhaust valve head and by short distance of flame travel.

2.11 Features of Different Types of Combustion Chamber Systems for S.I. Engine

Classification based on arrangement of valve:

1. L-head Engine

In this arrangement, all the valves are arranged in one line (except in case of V-8 engine) with the intake and exhaust valves arranged side by side. The combustion chamber and the cylinder are arranged in the form of inverted 'L'. All valves can be operated by a single crankshaft. Figure shows this arrangement.

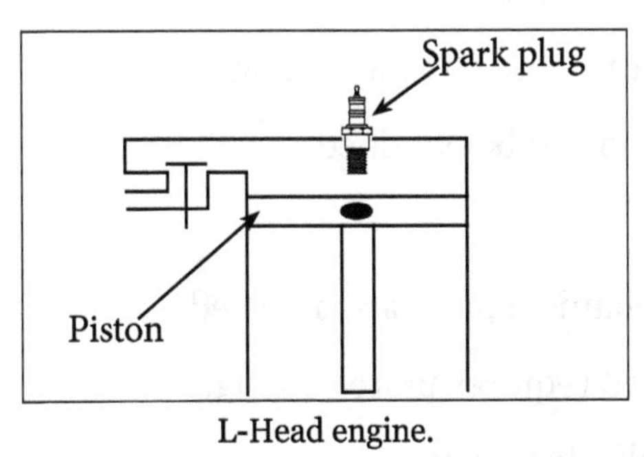

L-Head engine.

Advantages

1. One cam shaft is only required.
2. Height is reduced.
3. As the valves are arranged in one line, the removal of the cylinder is quite easy for servicing.
4. It is more dependable.

Disadvantages

1. More space for combustion chamber is required.
2. Knocking tendency is more than T-head engine.
3. Location of spark plug is difficult.
4. High compression ratio is not possible.

2. I-head Engine

In these engines, the cylinder head carries the valve. It is also called as overhead valve engine. In case of inline engines, the valves are arranged in a single row while the valves may be arranged in a single row or double row in each bank in case of V-engines. All the valves are actuated by single crankshaft.

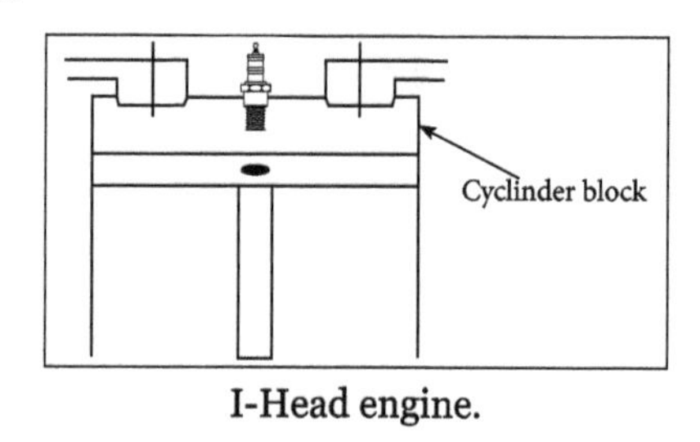

I-Head engine.

Advantages

1. A single camshaft actuates all valves.
2. Clearance volume is less. Hence, compression ratio can be increased considerably.
3. The spark plug can be located at center.
4. Smooth operation can be obtained.

Disadvantages

1. More value mechanism parts are involved.
2. The cylinder head requires more cooling.
3. It is more complicated design.
4. The size of the inlet and exhaust valve is limited.

3. T-Head Engine

It has the inlet valve on one side and the exhaust valve on the other side of the cylinder. Thus, two shafts are required to operate them. The combustion chamber and the cylinder form a letter 'T'. Generally, small engines are made with T-head arrangement.

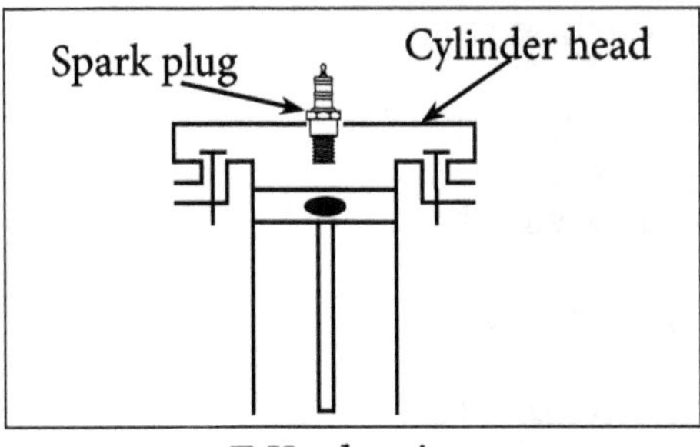

T-Head engine.

Disadvantages

1. Unequal temperature occurs in the cylinder.
2. More power is wasted in operating two camshafts.
3. Cost and weight are more.

4. F-Head Engine

In this arrangement, inlet valves are located in the cylinder head and exhaust valves by the sides of cylinders.

Advantages

1. More turbulence is possible.
2. More speed is possible.

Disadvantages

1. More space is required for the combustion chamber.
2. Location of spark plug is difficult.
3. Design of combustion chamber is difficult.

2.11.1 C.I. Engine Combustion Chambers: Open and Divided Type

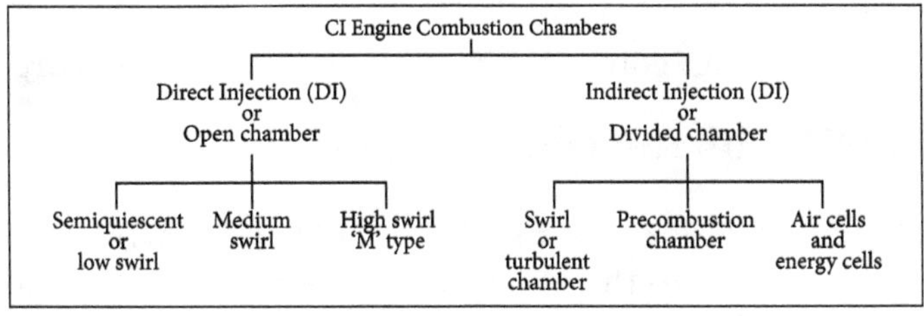

Classification of C.I engine.

C.I. engine combustion chambers are classified on the basis of method of generating swirl.

There are four types of combustion chambers:

1. Open combustion chambers.
2. Divided or turbulent swirl chambers.
3. Pre-combustion chambers.
4. Air cell combustion chambers.

The most important function of CI engine combustion chamber is to provide proper mixing of fuel and air in short time. In order to achieve this, an organized air movement called swirl is provided to produce high relative velocity between the fuel droplets and the air.

When the liquid fuel is injected into combustion chamber, the spray cone gets disturbed due to air motion and turbulence inside. The onset of combustion will cause an added turbulence that can be guided by the shape of the combustion chamber, makes it necessary to study the combustion design in detail.

C.I Engine combustion chambers are classified into two categories:

1. Open Injection (Di) Type: This type of combustion chamber is also called an Open combustion chamber. In this type the entire volume of combustion chamber is located in the main cylinder and the fuel is injected into this volume.

2. Indirect Injection (Idi) Type: In this type of combustion chambers, the combustion space is divided into two parts, one part in the main cylinder and the other part in the cylinder head. The fuel – injection is effected usually into the part of chamber located in the cylinder head.

These chambers are classified further into:

1. Swirl chamber in which compression swirl is generated.
2. Pre combustion chamber in which combustion swirl is induced.
3. Air cell in which both compression and combustion swirl are induced.

Direct Injection Chambers: Open Combustion Chambers

In this type the combustion space is essentially a single cavity with little restriction from one part of the chamber to the other and hence with no large difference in pressure between parts of the chamber during the combustion process.

In four-stroke engines with open combustion chambers, induction swirl is obtained either by careful formation of the air intake passages or by masking a portion of the circumference of the inlet valve whereas in two-stroke engines it is created by suitable form for the inlet ports.

These chambers mainly consist of space formed between a flat cylinder head and a cavity in the piston crown in different shapes. The fuel is injected directly into space. The injection nozzles used for this chamber are generally of multi hole type working at a relatively high pressure.

The main advantages of this type of chambers are:

1. Minimum heat loss during compression because of lower surface area to volume ratio and hence, better efficiency.
2. No cold starting problems.
3. Fine atomization because of multi hole nozzle.

The drawbacks of these combustion chambers are:

1. High fuel-injection pressure required and hence complex design of fuel- injection pump.
2. Necessity of accurate metering of fuel by the injection system, particularly for small engines.

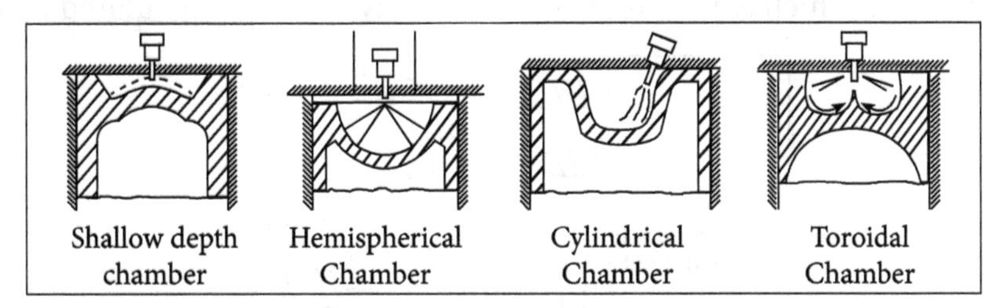

Shallow Depth Chamber

The depth of the cavity provided in the piston is quite small. This chamber is usually adopted for large engines running at low speeds. Since the cavity diameter is very large, the squish is negligible.

Hemispherical Chamber

This chamber also gives small squish. But the depth to diameter ratio for a cylindrical chamber can be varied to give any desired squish to give better performance.

Cylindrical Chamber

This design was attempted in recent diesel engines. This is a modification of the cylindrical chamber in the form of a truncated cone with base angle of 30°. The swirl was produced by masking the valve for nearly 1800 of circumference. Squish can also be varied by varying the depth.

Toroidal Chamber

The idea behind this shape is to provide a powerful squish along with the air movement, similar to that of the familiar smoke ring, within the toroidal chamber. Due to powerful squish the mask needed on inlet valve is small and there is better utilisation of oxygen. The cone angle of spray for this type of chamber is 150° to 160°.

In Direct Injection Chambers

A divided combustion chamber is defined as one in which the combustion space is divided into two or more distinct compartments connected by restricted passages. This creates considerable pressure differences between them during the combustion process.

Ricardo's Swirl Chamber

The swirl chamber consists of a spherical shaped chamber separated from the engine cylinder and located in the cylinder head. Into this chamber, about 50% of the air is transferred during the compression stroke.

A throat connects the chamber to the cylinder which enters the chamber in a tangential direction so that the air coming into this chamber is given a strong rotary movement inside the swirl chamber and after combustion, the products rush back into the cylinder through same throat at much higher velocity. This causes considerable heat loss to walls of the passage which can be reduced by employing a heat insulated passage.

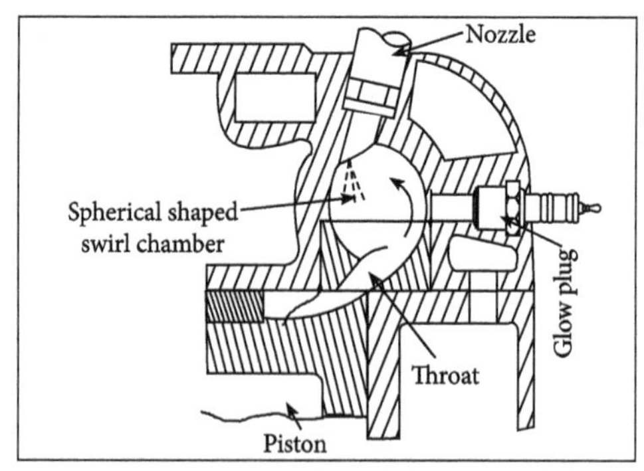

Ricardo's Swirl Chamber.

This type of combustion chamber finds its application where fuel quality is difficult to control, where reliability under adverse conditions is more important than fuel economy. The use of single hole of larger diameter for the fuel spray nozzle is often important consideration for the choice of swirl chamber engine.

Pre-Combustion Chamber

It consists of an anti chamber connected to the main chamber through a number of small holes. The pre-combustion chamber is located in the cylinder head and its volume accounts for about 40% of the total combustion, space. During the compression stroke the piston forces the air into the pre-combustion chamber.

The fuel is injected into the pre-chamber and the combustion is started. The resulting pressure rise forces the flaming droplets together with some air and their combustion products to rush out into the main cylinder at high velocity through the small holes. Thus it creates both strong secondary turbulence and distributes the flaming fuel droplets throughout the air in the main combustion chamber where bulk of combustion takes place.

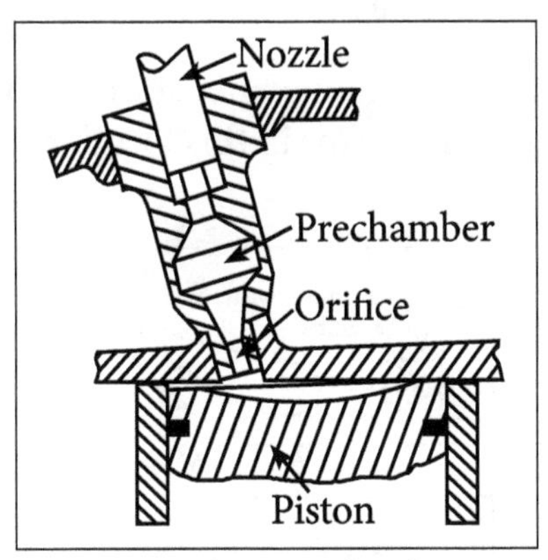

Pre Combustion Chamber.

About 80% of energy is released in main combustion chamber. The rate of pressure rise and the maximum pressure is lower compared to those in open type chamber. The initial shock occurs if combustion is limited to pre-combustion chamber only.

The pre-combustion chamber has multi fuel capability without any modification in the injection system because of the temperature of pre-chamber. The variation in the optimum injection timing for petrol and diesel operations is only 2 degree for this chamber compared to 8 to 10 degree in other chamber design.

Advantages

i. Due to short or practically no delay period for the fuel entering the main combustion space, tendency to knock is minimum and as such running is smooth.

ii. The combustion in the third stage is rapid.

iii. The fuel injection system design need not be critical. Because the mixing of fuel and air takes place in pre-chamber.

Disadvantages

i. The velocity of burning mixture is too high during the passage from pre-chambers, so the heat loss is very high. This causes reduction in the thermal efficiency, which can be offset by increasing the compression ratio.

ii. Cold starting will be difficult as the air loses heat to chamber walls during compression.

Energy Cell

The 'energy cell' is more complex than the pre- combustion chamber. As the piston moves up on the compression stroke, some of the air is forced into the major and minor chambers of the energy cell. When the fuel is injected through the pintle type nozzle, part of the fuel passes across the main combustion chamber and enters the minor cell, where it is mixed with the entering air.

Combustion first commences in the main combustion chamber where the temperature is higher, but the rate of burning is slower in this location, due to insufficient mixing of the fuel and air.

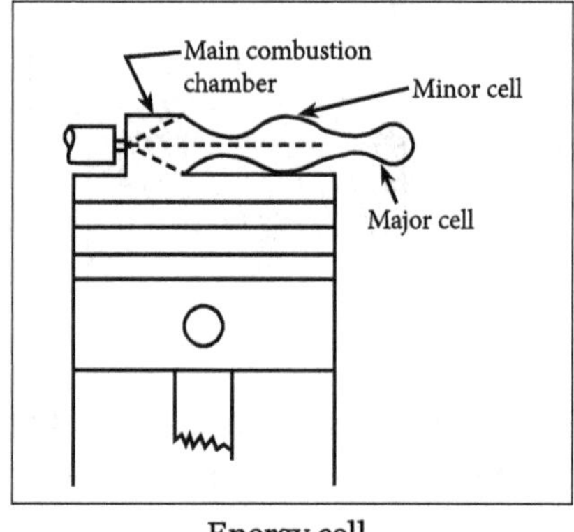

Energy cell.

The burning in the minor cell is slower at the start, but due to better mixing, progresses at a more rapid rate. The pressure built up in the minor cell, therefore, force the burning gases out into the main chamber, thereby creating added turbulence and producing better combustion in the chamber. In mean time, pressure is built up in the major cell which then prolongs the action of the jet stream entering the main chamber, thus continuing to induce turbulence in the main chamber.

2.12 Air Swirl Turbulence: M-Type Combustion Chamber

Spiral intake ports produce a high speed rotary air motion in the cylinder during the induction stroke. Here, a single coarse spray is injected from a pintle nozzle in the direction of the air swirl and tangential to the spherical wall of the combustion chamber in the piston. The strikes against the wall of the spherical combustion chamber where it spreads to form a thin film which will evaporate under controlled conditions.

The air swirl in the spherically shaped combustion chamber is quite high which sweeps over the fuel film, peeling it from the wall layer by layer for progressive and complete combustion. The flame spirals slowly inwards and around the bowl, with the rate of combustion controlled by the rate of vaporization.

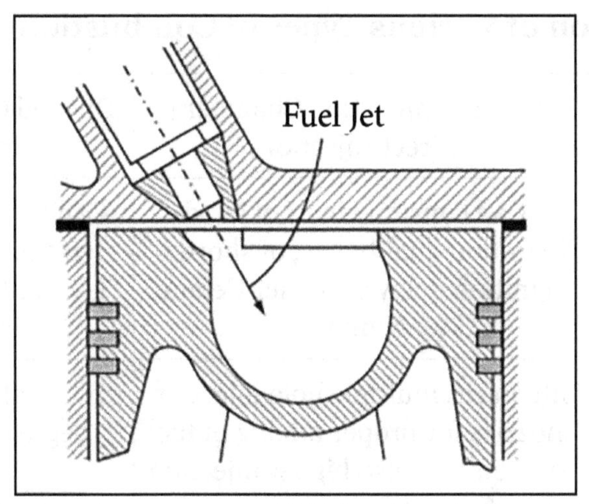

Direct injection bowl-in-piston chamber with high swirl and a single hole nozzle, M-type.

Figure Shows the 'M' MAN Meurer wall burning combustion system, optimized to give greater swirl and mixing for complete combustion. This combustion system was developed around 1954 by the Maschin-enfabrik Augsburg Numburg (MAN) AG of Germany for small, high speed engines.

It has single-hole fuel-injection, so oriented that most of the fuel is deposited on the piston bowl walls. In practice, this engine gives good performance even with fuels of exceedingly poor ignition quality. Its fuel economy appears to be extremely good for an engine of small size. Because of the vaporization and mixing processes, the 'M' engine is ideally suited as a multi fuel engine.

Advantages

i. Low rates of pressure rise, low peak pressure.

ii. Low smoke level.

iii. Ability to operate on a wide range of liquid fuels (multi-fuel capability).

iv. No combustion noise is reported even for 80-octane petrol.

Disadvantages

i. Since fuel vaporization depends upon the surface temperature of the combustion chamber, cold starting requires certain aids.

ii. Some white smoke, diesel and high hydrocarbon emission may occur at starting and idling conditions.

iii. Volumetric efficiency is low.

2.12.1 Comparison of Various Types of Combustion Chambers

SL.NO	Aspects	Open Combustion Chamber (Direct Injection)	Divided Combustion Chamber (Indirect Injection)
1	Fuel Used	Can consume fuels of good ignition quality. i.e., of shorter ignition delay or higher Cetane Number	Can consume fuels of poor ignition quality i.e. of longer ignition delay or lower Cetane Number
2	Type of injection nozzles used	Requires multiple hole injection nozzles for proper mixing of fuel and air and also higher injection pressures	It can also tolerate greater degree of nozzle fouling
3	Sensitivity to fuel	Sensitive	Insensitive
4	Mixing of fuel and air	Mixing of fuel and air is not so efficient and thus high fuel air ratios are not feasible without smoke	Ability to use higher fuel ratio without smoke due to proper mixing and consequent high air utilization factor
5	Cylinder construction	Simple	More expensive cylinder construction
6	Starting	Easy cold starting	Difficult to cold start because of greater heat loss through throat
7	Thermal efficiency	Has more open combustion chambers are thermally more efficient	Thermal efficiency is lower due to throttling in throat areas leading to pressure losses and heat losses.

2.13 Super Charging and Scavenging: Thermodynamic Cycles of Supercharging

Following figure shows the thermodynamics cycle of a supercharged I.C Engine on the p-v diagram for an otto cycle:

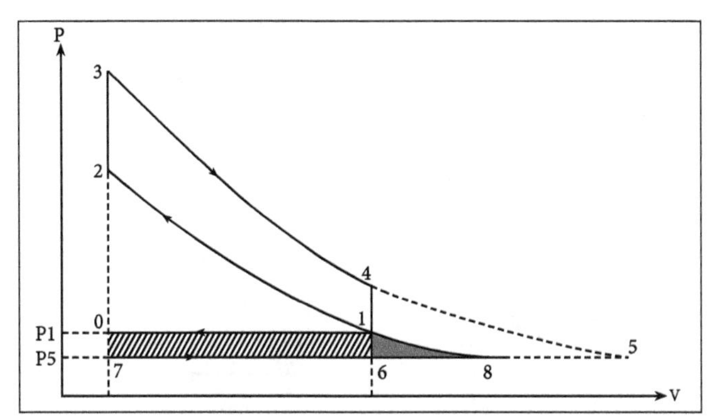

Thermodynamic cycle of supercharged engine on p-v diagram form ideal otto cycle.

1. The pressure p1, represents the supercharging pressure and p5, is the exhaust pressure.
2. The thermodynamic cycle, consists of the following processes:

i. 0-1: Admission of air at the supercharging pressure (which is greater than atmospheric pressure).

ii. 1-2: Isentropic compression.

iii. 2-3: Heat addition at constant volume (for diesel cycle, this will be replaced by a constant pressure process, representing heat addition at constant pressure.

iv. 3-4: Isentropic expansion.

3. 4-1-6: Heat rejection at constant volume (blowdown to atmospheric pressure).
4. 6-7: Driving out exhaust at constant atmospheric pressure.

The thermodynamic cycle for the supercharger consists of the following processes:

1. 7-6. Admission of air at atmospheric pressure.
2. 8-1. isentropic compression to pressure.
3. 1-0. Delivery of supercharger air, at a constant pressure.

Area 8-6-7-0-1-8 represents the supercharger wank (mechanically driven) in supplying air at a power. While the area 1-2-3-4-1, is the output of the engine. Area 0-1-6-7-0 represents the gain in work during the gas. That leaves process due to supercharging. Thus a part of the supercharger work is recovered. However, the work represented by the area 1-6-8-1 cannot be recovered and represents a loss of work.

This improvement in combustion allows a poor quality fuel to be used in a diesel engine and it is also not sensitive to the type of fuel used. The increase in intake temperature reduces volumetric and thermal efficiency but increase in density due to pressure compensates for this and inter cooling is not necessary except for highly supercharged engines.

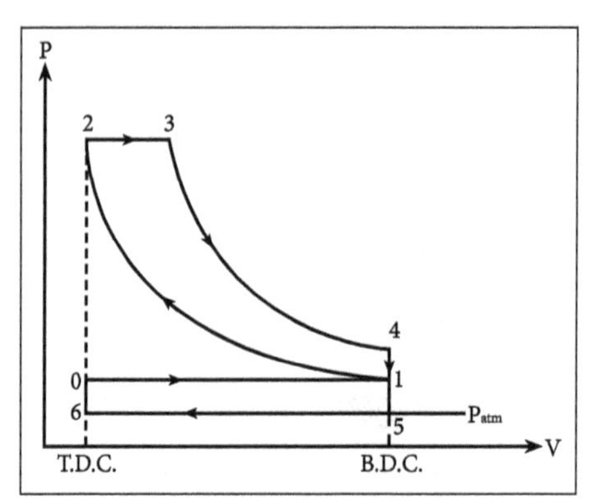

Supercharging C.I engine.

If a supercharger engine is supercharged it will increase the reliability and durability of the engine due to smoother combustion and lower exhaust temperatures. The degree of supercharging is limited by thermal and mechanical load on the engine and strongly depends on the type of supercharger used and design of the engine.

2.14 Effect of Supercharging and Efficiency of Supercharged Engines

Effect of Supercharging Barging on Performance of the Engine

1. The 'power output' of a supercharged engine is higher than its naturally aspirated counterpart.
2. The 'mechanical efficiency's of supercharged engines are slightly better than the naturally aspirated engines.
3. In spite of better mixing and combustion due to reduced delay a mechanically supercharged otto engine almost always have 'specific fuel consumption' higher than a naturally aspirated engine.

Effect of Supercharging on Mechanical Efficiency

Due to the packing of more charge in the engine cylinder during supercharging, the gas load is increased which needs large bearings and robust components. Heavy load on the bearings increases the friction force.

However, the increase in the brake mean effective pressure is much more than the increase in frictional forces. Therefore the mechanical efficiency of a supercharged engine is improved as compared to a supercharged engine. This mechanical efficiency increases with the increase in the degree of supercharging, but reduces with the increase in engine speed.

Effect of Supercharging on Fuel Consumption

The power required to drive the supercharger varies with different arrangements. If the super-charger is directly driven by the engine. then extra fuel is supplied to compensate for the loss of power which is required to drive the compressor. If a gasoline engine runs at periodic load, then the throttle valve is partially closed. In such conditions the effect of supercharging is neutralized. This results in a greater loss and the specific fuel consumption is increased.

A highly supercharged gasoline engine needs a very rich mixture to avoid knock and pre-ignition. This also gives rise to higher specific fuel consumption. The specific fuel consumption in a supercharged diesel engine is lesser than the tin-supercharged diesel engine and is due to better combustion and increased mechanical efficiency.

Exhaust-driven superchargers do not require any power from the engine. On the other hand, a part of the exhaust energy is utilized for work which gives about 5% better thermal efficien-cy at full load when the throttle valve is fully opened. This lowers specific fuel consumption.

2.14.1 Methods of Super Charging, Supercharging and Scavenging of 2-Stroke Engines

Turbo Chargers and Superchargers

Superchargers are driven by a gas turbine which derives from the engine exhaust gas. This type of supercharger is called turbocharger. Superchargers are only the pressure boosting device which supplies air or air fuel mixture to engine with high pressure.

Purpose of Turbo Charger

1. To reduce weight per horse power of the engine as required in air plane engine.
2. To reduce the space occupied by the engine as required in marine engine.
3. To maintain the power of a reciprocating IC engines even at high altitude where less oxygen is available for combustion.

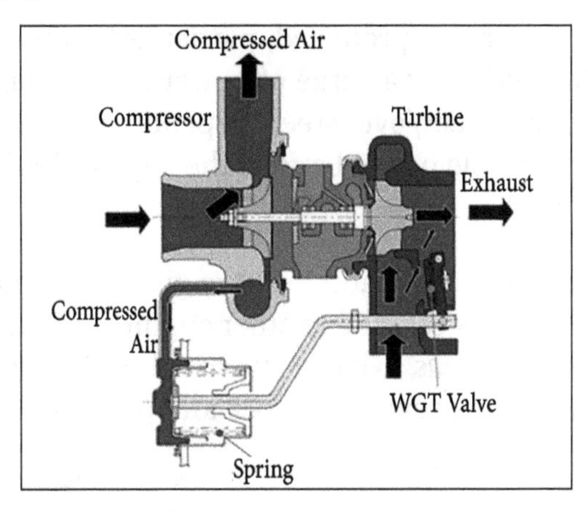

The waste gate valve is installed to prevent turbo breakage from excess intake pressure and exhaust pressure in high engine speed. A boost pressure follows engine speed and amount of stroke follows pressure. A spring is selected with suitable spring constant to operate actuator with the above values.

Modifications to be Done in an Engine to Make It Suitable For Super Charging

Supercharging results in the increased output of a naturally-aspirated engine. The following engine modifications are recommended for trouble free supercharging:

i. The valve overlap period should be increased to allow complete scavenging of the exhaust gases from the clearance space.

ii. The compression ratio should be reduced in order to increase the clearance volume. The effect of this is to reduce mechanical and thermal loading on the engine.

iii. In the SI engine the spark timing should be retarded in order to reduce the maximum pressure of the cycle.

iv. In the CI engine the injection system should be modified to supply increased amount of fuel. This will require a larger nozzle area than that in the naturally aspirated engine.

v. In turbocharged engines the exhaust valve should open a little earlier to supply more energy to the turbocharger and the exhaust manifold should be insulated to reduce heat losses.

Limitations of Turbo Charging

i. Turbo lag, especially on large turbos. A large turbo may give more peak power, but can take more time to spool up.

ii. Drive ability may be compromised, particularly when the boost threshold is approached and suddenly a surge of power is too much for the tyres to cope with, causing under steer/over steer (depending on which wheels are driven). This reduces the usable power band of the engine and leads to more wear and tear on the train.

iii. Turbochargers are costly to add to NA engines and add complexity. Adding a turbo can often cause a cascade of other engine modifications to cope with the increased power, such as exhaust manifold, inter cooler, gauges, plumbing lubrication and possibly even the block and pistons.

Types of Superchargers

Superchargers are air pumps used for the purpose of delivering large quantity of air to the engine inlet. i.e., to supercharge the engines. They can be fitted to any engines petrol or diesel, two-stroke or four stroke. They are driven by tapping the engine power. Several types of super chargers viz:

1. Piston type.
2. Centrifugal type.
3. Vane type.
4. Root's type.
5. Turbo type.

Centrifugal Type Supercharger

This supercharger delivers air under the centrifugal action. It consists of following main components:

1. Impeller with an eye.
2. Diffuser or roster chamber.
3. Volute Scroll casing.
4. Aero foiled blades on impeller and diffuser.

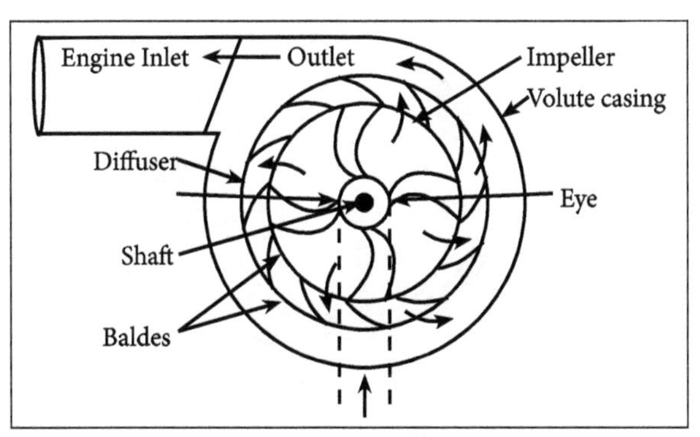

Centrifugal Type Supercharger.

When supercharger shaft is run by the engine, the impeller also rotates with it as it is mounted on this shaft. The rotations induces various, inside the casing which causes the air to fill into it. Therefore, the air rushes in through the eye. The air flows over the impeller blades, passes over the diffuser blades and then to the volute casing.

The propeller and the diffuser are circular members while the volute casing is of variable cross-sectional area. When the air flows over blades of the impeller, a part of its

K.E is recovered and is converted into pressure. The Diffuser is provided to transform the K.E into P.E. The advantage Is that it is compact in construction, occupies less space and the auto vehicle is capable of pumping enough air.

Vane Type Super Charger

This is a positive displacement type supercharger in which a rotor mounted eccentrically or the shaft moves within the housing.

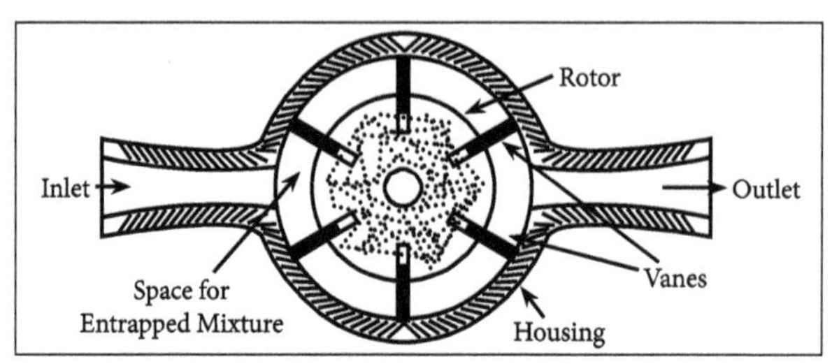

Vane type super charger

The rotor carries vanes which move in and within the housing against the spring force. The pumping of the air is accomplished by centrifugal actions. Its construction is similar to the race type oil pump. The construction of housing is shaped such that the space between it and the rotor decreases continuously from inlet to the outlet.

The vanes always remains in contact with the inner side of the housing. Although suitable for use at lower speeds, such superchargers suffer from the problem of wear of the vanes. Hence, they are not much used.

Turbocharger

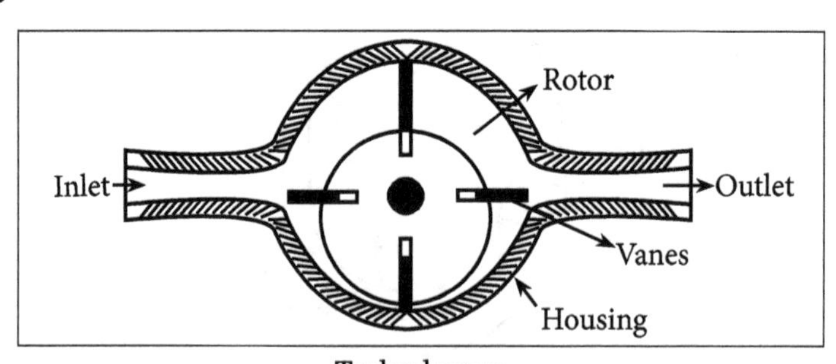

Turbocharger

This a positive displacement type super charger in which a rotor mounted eccentrically or the shaft moves within the housing. The rotor carries vanes which move in and within the housing against the spring force. The pumping of the air is accomplished by centrifuge actions.

3

An Overview of Lubricating Systems, Engine Emission, Gas Turbines and Air Craft Propulsion

3.1 Testing and Performances: Power, Fuel and Air Measurement Methods

A power is defined as the rate of doing work and is equal to the product of force and linear velocity or the product of torque and angular velocity. Thus, the measurement of power involves the measurement of force as well as speed. The force or torque is measured with the help of a dynamometer and the speed by a tachometer.

$$b_p = \frac{2\pi NT}{60}$$

Where,

T -Torque in $N-m$.

N -Rotational speed in revolutions per minute.

The total power developed by combustion of fuel in the combustion chamber is more than the b_p and is called indicated power (i_p). It is given by,

$$IP = \frac{P_m LANk}{60}$$

Where,

P_m =Mean effective pressure, N/m^2

L=Length of the stroke, m,

A=Area of the piston, m^2,

N=Rotational speed of the engine, rpm (It is N/2 for four stroke engine).

k=Number of cylinders.

It is the power developed in the cylinder and forms the basis of evaluation of combustion efficiency or the heat release in the cylinder. Thus, we see that for a given engine the power output can be measured in terms of mean effective pressure.

Measurement of Brake Power

The brake power measurement involves the determination of the torque and the angular speed of the engine output shaft. The torque measuring device is known as dynamometer. Dynamometers can be broadly classified into two main types, power absorption dynamometers and transmission dynamometer.

A rotor driven by the engine under test is electrically, hydraulically or magnetically coupled to a stator. For every revolution of the shaft, the rotor periphery moves through a distance $2\pi r$ against the coupling force F. Hence, the work done per revolution is,

$$W = 2\pi RF$$

The external moment or torque is equal to $S \times L$ where, S is the scale reading and L is the arm length. This moment balances the turning moment $S \times L$, i.e.

$$S \times L = R \times F$$

$$\because \text{Work done / revolution} = 2\pi SL, \text{ Work done / minute} = 2\pi SLN$$

Where, N is rpm. Hence, power is given by,

Brake power $P = 2\pi NT$.

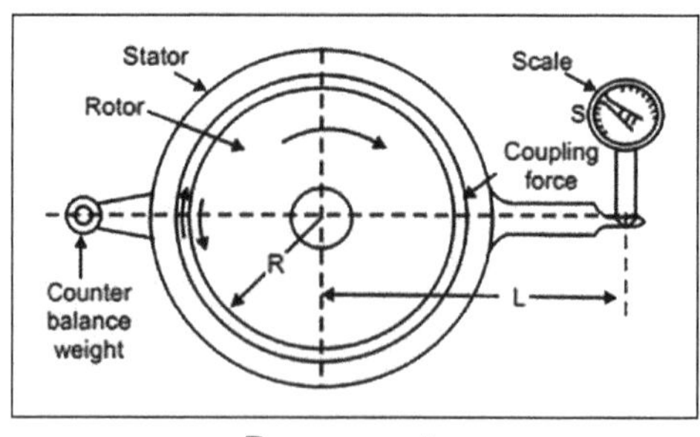

Dynamometer

Prony Brake

One of the simplest methods of measuring brake power (output) is to attempt to stop the engine by means of a brake on the flywheel and measure the weight which an arm attached to the brake will support, as it tries to rotate with the flywheel. This system is known as the prony brake and forms its use. It works on the principle of converting power into heat by dry friction.

It consists of wooden block mounted on a flexible rope or band, the wooden block when pressed into contact with the rotating drum takes the engine torque and the power is dissipated in frictional resistance. Spring-loaded bolts are provided to tighten the wooden block and hence increase the friction. The whole of the power absorbed is converted into heat and hence this type of dynamometer must the cooled. The brake horsepower is given by,

$$BP = 2\pi NT$$

Where, $T = W \times l$

W being the weight applied at a radius l.

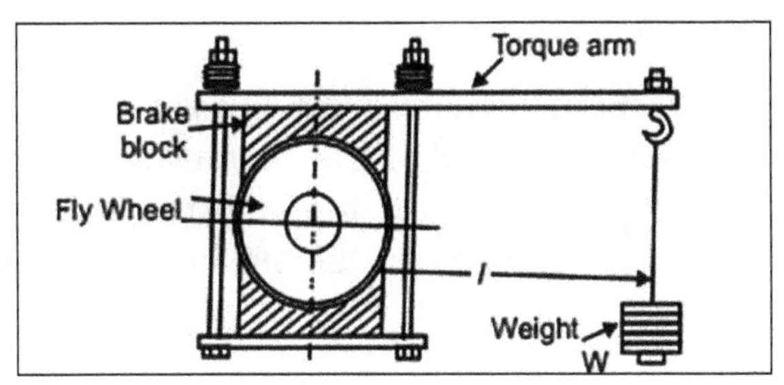

Prony Brake.

Rope Brake

The rope brake is another simple device for measuring brake power of an engine. It consists of a number of turns of rope wound around the rotating drum attached to the output shaft. One side of the rope is connected to a spring balance and the other to a loading device. The power is absorbed in friction between the rope and the drum. The drum therefore requires cooling.

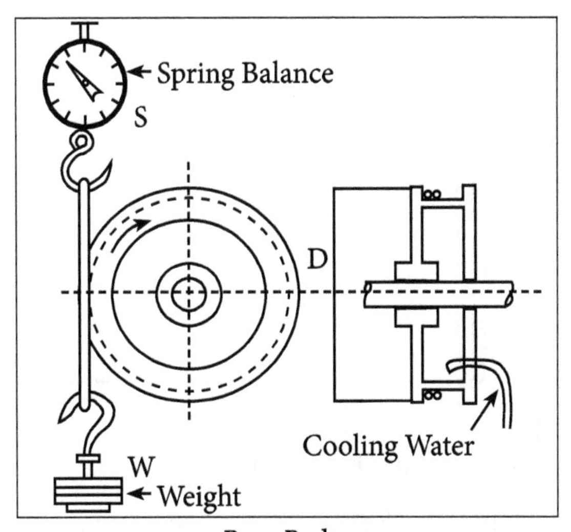

Rope Brake.

Rope brake is cheap and easily constructed but not a very accurate method because of changes in the friction coefficient of the rope with temperature.

The b_p is given by,

$$b_p = \pi DN(W - S)$$

Where,

D -Brake drum diameter.

W -Weight in Newton.

S -Spring scale reading.

Fuel-Air Measurement

Fuel-air ratio (F/A) is the ratio of the mass of fuel to the mass of air in the fuel-air mixture. Air-fuel ratio (A/F) is reciprocal of fuel-air ratio. Fuel-air ratio of the mixture affects the combustion phenomenon as it determines the flame propagation velocity, the heat release in the combustion chamber, the maximum temperature and the completeness of combustion.

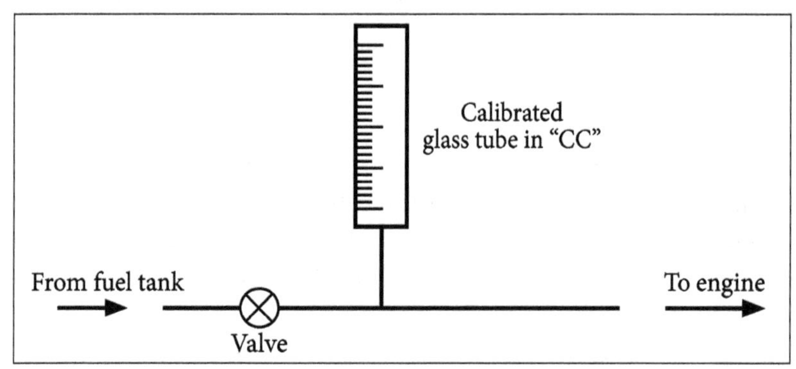

Fuel flow measurement.

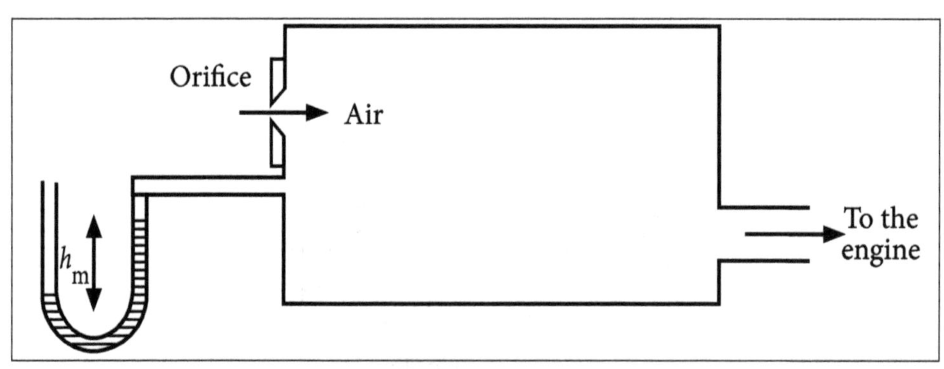

Air flow measurement.

Relative fuel-air ratio is defined as the ratio of the actual fuel-air ratio to that of the stoichiometric fuel-air ratio required to burn the fuel supplied. Stoichiometric fuel-air ratio is equal to one and in which case fuel is completely burned due to minimum quantity of air supplied.

$$\text{Relative fuel-air ratio, } F_R = \frac{\text{Actual fuel} - \text{Air ratio}}{\text{Stoichiometric fuel} - \text{Air ratio}}$$

Air Measurement of Method

In IC engines, the satisfactory measurement of air consumption is quite difficult because the flow is pulsating, due to the cyclic nature of the engine and because the air is a compressible fluid. Therefore, the simple method of using an orifice in the induction pipe is not satisfactory since the reading will be pulsating and unreliable.

All kinetic flow-inferring systems such as nozzles, orifices and ventures have a square law relationship between flow rate and differential pressure which gives rise to severe errors on unsteady flow. Pulsation produced errors are roughly inversely proportional to the pressure across the orifice for a given set of flow conditions.

The various methods and meters used for flow measurement include:

1. Air box method.
2. Viscous-flow air meter.

3.1.1 Performance Characteristic Curves of SI and CI Engines

The word performance for an engine is generally used for designating the relationship between power, speed and fuel consumption. In variable speed engines (like automobile engines), the rated power at a particular speed does not provide enough information. Under such situations, the performance curves help to obtain necessary information. Typical performance curves for internal combustion engines (both petrol and diesel engines) used in automobiles are shown in figure:

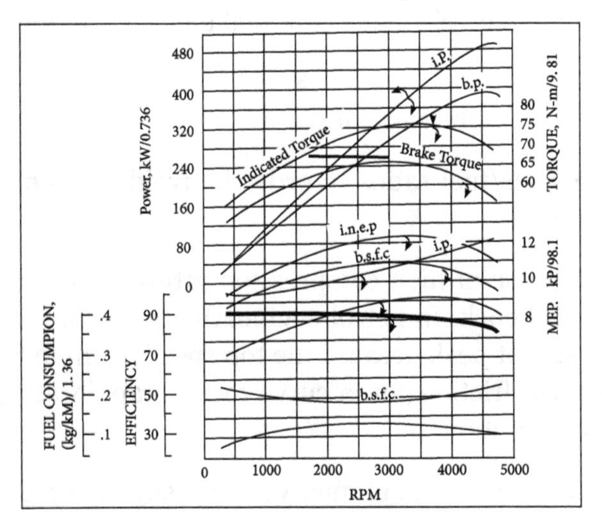

Typical performance curves of automotive SI engine.

Figure above shows performance curves for high output, V-8, multiple carburettor (3 dual-throat carburettor) automotive petrol engine with 7 x 10^{-3} m^3 displacement. Figure shows below the performance curves for typical automotive CI engine having 6-cylinders (110 mm×135 mm) and compression ratio 15:1 that uses 50 cetane fuel.

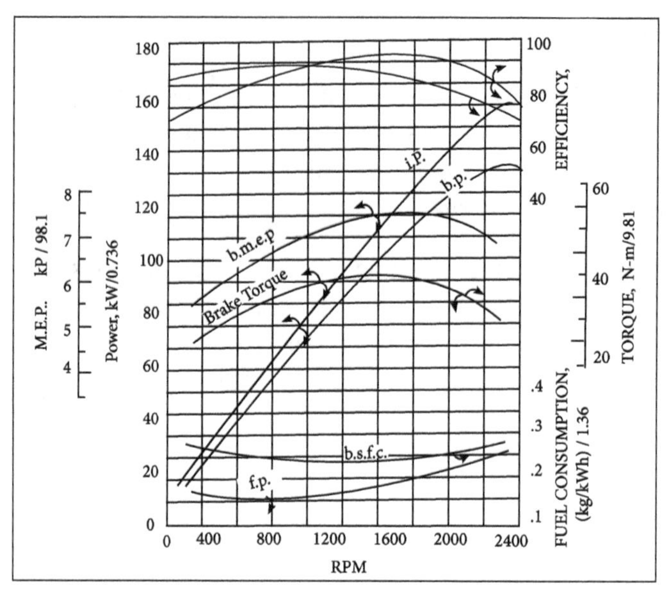

Typical performance curves of automotive CI engine.

The two figures reveal that in diesel engines, fuel consumption per kWh is less and marked at the usual range of part-load operation. The torque for diesel engine remains fairly uniform over a wider range of operating speeds than for petrol engine. This results in better top gear performance, as the engine is more flexible over a wider speed range. Moreover, a high value of the torque at lower engine speeds in diesel engine enables the vehicle to run much more slowly on top gear.

The CI engine has an appreciable higher thermal efficiency than the petrol engine because,

1. It has a higher compression ratio.
2. It uses higher air/fuel ratios in order to avoid incomplete combustion and smoky exhaust.

On the other hand, the use of high compression ratios in petrol engines creates the problem of combustion knock. In petrol engine, the brake power curve takes a peak, but in CI engine it does not peak because the top speed is limited due to their heavier reciprocating masses. The friction power curve goes up rapidly at higher speeds in both cases as it includes fluid friction.

It is a common practice to give the automotive diesel engine three ratings:

1. The maximum rating for short intervals of operation.
2. The rated output for larger period of operation than the former.
3. The continuous output rating for operation with no time limit.

3.2 Variables Affecting Performance and Methods to Improve Engine Performance

Volumetric Efficiency

Volumetric efficiency of an engine is an indication of the measure of the degree to which the engine fills its swept volume. It is defined as the ratio of the mass of air inducted into the engine cylinder during the suction stroke to the mass of the air corresponding to the swept volume of the engine at atmospheric pressure and temperature.

$$\eta_v = \frac{\text{Mass of charge actually sucked in}}{\text{Mass of charge corresponding to the cylinder intake P and T conditions}}$$

Thermal Efficiency

Thermal efficiency of an engine is defined as the ratio of the output to that of the chemical energy input in the form of fuel supply. It may be based on brake or indicated output.

$$\text{Brake thermal efficiency} = \frac{b_p}{m_f \times C_v}$$

Where, $C_v = \text{Calorific value of fuel, kJ / kg}$, and

$m_f = \text{Mass of fuel supplied, kg / sec}$.

Heat Balance Sheet

Engine heat transfer analysis is required for the estimation of material temperature limits, lubricant performance limits, emissions and knock. As the combustion process in an IC engine is not continuous, the component temperatures are much less than the peak combustion temperatures. But the temperatures of certain critical areas need to be kept below material design limits as aluminium alloys and iron begin to melt at temperatures greater than 540°C and 1500°C respectively.

Temperature difference around the cylinder bore will cause bore distortion and increase oil consumption and piston wear. Cooling of the engine is also required to prevent knock in spark ignition engines.

Engine	Brake Load Efficiency	Loss To Cooling Water	Loss To Exhaust	Other Losses, Friction Losses And Loss Due To Incomplete Combustion
Gasoline engine or spark ignition	21 % to 28 %	12 % to 27 %	30 % to 55 %	3 % to 55 % (incomplete combustion loss 0 to 45 %)
Diesel engine or compression ignition engine	29 % to 42 %	15 % to 35 %	25 % to 45 %	21 % to 0 % (incomplete combustion loss 0 to 5 %)

Heat transfer analysis in exhaust pipe is also an important factor for emissions and satisfactory catalytic converter performance occurs above some temperature. The temperature required for the catalyzed oxidation of hydrocarbon and carbon monoxide emissions is about 230°C and hence exhaust temperatures are to be maintained above this limit for better converter performance.

Energy balance of an IC engine can be obtained through experiments by measuring the quantities of heat rejected to water and to the ambient air. Flow meters and thermocouples measure the water and exhaust gas flow and inlet and outlet temperatures.

An energy balance applied to the water coolant through the engine and exhaust gas yields:

$$HLCW = m_w C_{pw} \left(T_{wo} - T_{wi}\right)$$

$$HLEG = m_{eg} C_{pg} \left(T_{ego} - T_{egi}\right)$$

Determining the heat loss to the ambient air as unaccounted loss is more complicated and hence from the difference of fuel power and the sum of heat lost to cooling water and exhaust gas.

Unaccounted loss = Fuel power- $(HLCW + HLEG)$

$\text{Fuel Power} = \text{mass flow rate of fuel} \times \text{calorific value of fuel.}$

The mass flow rate of exhaust emissions is given by,

$$m_{ex} = m_{air} + m_{fuel}$$

Where,

$HLCW = \text{Heat loss to cooling water.}$

$HLEG = \text{Heat loss to exhaust gas.}$

$T_{wo} = \text{Cooling water exit temperature.}$

$T_{wi} = \text{Cooling water inlet temperature.}$

T_{ego} = Exhaust gas outlet temperature.

$T_{egi} = \text{Exhaust gas inlet temperature (Room air temperature).}$

$C_{pw}, C_{pg} = \text{Specific heat of water and exhaust gas.}$

Problems

1. A twin-cylinder two-stroke engine has a swept volume of 150 cm3. The maximum power output is 19 kW at 11000 rpm, bsfc is 0.11 kg/MJ and the air/fuel ratio is 12. If

ambient test conditions were 100C and 1.03 bar and the fuel has a calorific value of 44 MJ/kg. Let us calculate the bmep, overall efficiency and the volumetric efficiency.

Solution:

Given:

Volume -150cm^3

P= 19kw

Rpm= 11000

bsfc= 0.11kg / MJ

Ratio = 12

$$\text{bsfc} = \frac{\text{Fuel consumption}}{\text{BP}}$$

$$0.11 \times 10^{-6} = \frac{\text{Fuel consumption}}{19 \times 10^3 \text{ J/s}}$$

$$\text{Fuel consumption} = 0.11 \times 10^{-6} \times 19 \times 10^3 \left(\text{J/s kg/J}\right)$$

$$= 2.09 \times 10 \quad \text{kg/s}$$

Brake mean effective pressure,

$$\text{bmep} = \frac{\text{BP}}{\left(\pi d^2/4\right)\left(\text{In k}\right)}$$

$$\left(\pi d^2 l/4\right) \times n \times K = \frac{150 \times 10^{-6} \times 11000 \times 1}{60} \text{m}^3/\text{sec}$$

$$= 0.0275 \text{m}^3/\text{sec}$$

$$\text{bmep} = \frac{19 \times 10^3}{0.0275} = 690909 \text{N/m}^2$$

$$\text{bmep} = 6.9 \text{bar}$$

Overall efficiency or brake thermal efficiency,

$$\eta_{\text{overall}} = \frac{\text{Brake power}}{m_t \times \text{CV}}$$

$$\eta_{overall} = \frac{19\times10^3}{2.09\times10^{-3}\times44\times10^6}\times100 = 20.66\%$$

Volumetric efficiency, $\eta_{vol} = \frac{\text{Actual volume of air intake}}{\text{Theoretical volume flow rate}} = \frac{V_3}{V_3}$

$$\text{Air flow} = 12\times2.09\times10^{-3}\ \text{kg/sec.}$$

$$m_3 = 0.02508\ \text{kg/sec.}$$

$$PV = mRT$$

$$P_1V_1 = m_3RT_1$$

$$V_1 = \frac{0.02508\times(287)\times(283)}{1.03\times10^5}$$

$$V_1 = 0.01977\ \text{m}^3/\text{sec.}$$

$$\eta_{vol} = \frac{0.01977}{150\times10^{-6}\times\frac{11000}{60}} = 0.719 = 71.9\%$$

2. The Morse test on a four-stroke four-cylinder petrol engine having bore of the cylinder = 80 mm, stroke length = 120 mm, clearance volume = 8 x 10^4 mm^3. Fuel consumption = 4.74 kg/hr, calorific value of fuel = 44,000 kJ/kg, BP with all cylinders working = 14.5 kW, BP with Cylinder 1 cut-off =10 kW, BP with Cylinder 2 cut-off = 10.2 kW, BP with Cylinder 3 cut-off=10.1 kW and BP with Cylinder 4 cut-off = 10.3 kW. Let us estimate mechanical, indicated, brake thermal and relative efficiencies.

Solution:

Given:

$$L = 120\text{mm}$$
$$\text{Volume} = 8\times10^4\,\text{mm}^3$$
$$\text{COF} = 44000\,\text{kj/kg}$$
$$P = 1405\ \text{kW}$$

1 cut-off =10 kW, BP with Cylinder 2 cut-off = 10.2 kW, BP with Cylinder 3 cut-off = 10.1 kW and BP with Cylinder 4 cut-off = 10.3 kW.

$$IP_{engine} = IP_1 + IP_2 + IP_3 + IP_4$$

$$=\left(BP_{1,2,3,4}-BP_{2,3,4}\right)+\left(BP_{1,2,3,4}-BP_{1,3,4}\right)$$

$$+\left(BP_{1,2,3,4}-BP_{1,2,4}\right)+\left(BP_{1,2,3,4}-BP_{1,2,3}\right)$$

$$=4BP_{1,2,3,4}-\left(BP_{2,3,4}+BP_{1,3,4}+BP_{1,2,4}+BP_{1,2,3}\right)$$

$$=4(14.5)-(10+10.2+10.1+10.3)$$

$$IP_{engine}=17.4\,kw$$

Fuel Power, $FP = FC \times CV$

$$=\frac{4.47\times 44{,}000}{3600}=57.93\,kw$$

$$BP = 14.5\ kw$$

Air standard efficiency, $\eta = \frac{1}{R^{\gamma-1}}$

Compression ratio, $R=\frac{V_c+V_s}{V_c}$

$$=\frac{8\times 10^4+\frac{\pi}{4}\times 80^2\times 120}{8\times 10^4}=8.54$$

$$\text{Air standard efficiency}=1-\frac{1}{8.54^{0.4}}=0.576=57.6\%$$

$$\text{Mechanical efficiency}=\frac{BP}{IP}=\frac{14.5}{17.4}\times 100=83.33\%$$

$$\text{Indicated thermal efficiency}=\frac{IP}{FP}=\frac{17.4}{57.93}\times 100=30.04\%$$

$$\text{Brake thermal efficiency}=\frac{BP}{FP}=\frac{14.5}{57.93}\times 100=25.03\%$$

$$\text{Relative efficiency}=\frac{\text{Indicated thermal efficiency}}{\text{Air std. efficiency}}=\frac{0.3004}{0.576}=0.522$$

$$=52.2\%$$

3. A test was conducted on a single-cylinder two-stroke engine where net brake load on engine is 600 N, speed is 400 rpm, mean effective pressure is 3.5 bar, fuel consumption is 4.5 kg/hr, cooling water flow rate is 500 kg/hr, water inlet and outlet temperatures are 30°C and 55°C, respectively, air fuel ratio is 26, air temperature is 30 °C, exhaust gas temperature is 320 °C, cylinder bore is 21 cm, stroke length is 25 cm, brake drum diameter is 98 cm, calorific value of fuel is 44,000 kJ/kg. Let us estimate the mechanical efficiency, indicated thermal efficiency and draw up a heat balance chart on percentage basis.

Solution:

Given:

Load = 600N

Speed = 400rpm

P = 3.5 bar

$$\text{Fuel power} = FC \times CV$$

$$= \frac{4.5 \times 44{,}000}{3600} = 55\,\text{kw}$$

$$\text{Brake power} = \frac{2\pi NT}{60{,}000}$$

$$= \frac{2\pi(400)(600)(0.49)}{60{,}000} = 12.32\,\text{kw}$$

$$\text{Indicated power, } IP = \frac{P_{mef}.l.A.n.K}{60{,}000}$$

$$= \frac{350 \times 10^5 \times 0.25 \times \frac{\pi}{4}(0.21)^2 \times 400 \times 1}{60{,}000} = 20.2\ \text{kW}$$

$$\text{Mechanical efficiency} = \frac{BP}{IP} = \frac{12.32}{20.2} = 0.6099 = 60.99\%$$

$$\text{Indicated thermal efficiency} = \frac{IP}{FP} = \frac{20.2}{55} = 0.3673 = 36.73\%$$

Heat Balance:

$FP = 55$ kw

$BP = 12.32$ kW

$$HLCW = m_{cw} \times C_{pcw} \times \Delta T_{cw}$$

$$= \frac{500 \times 4.18 \times (55-30)}{3600} = 14.51 \, kw$$

$$HLEG = m_{eg} \times CP_{eg} \times \Delta T_{eg}$$

$$= (m_{fuel} + m_{air}) \times (1.005) \times (320-30)$$

$$= \frac{(26+1)(4.5) \times 1.005 \times 290}{3600} = 9.84 \, kw$$

$$\text{Unaccounted losses} = FP - (BP + HLCW + HLEG)$$

$$= 55 - (12.32 + 14.51 + 9.84)$$

$$= 18.33 \, kw$$

The heat balance is given in table:

Heat Balance:

Fuel power	BP	HLEG	HLCW	Unaccounted losses
55 kW	12.32 kW	9.84 kW	14.51 kW	18.33 kW
HE: 100%	22.4 %	17.89 %	26.38 %	33.33 %

3.3 Cooling and Lubricating Systems

When the air-fuel mixture is ignited combustion takes place at about 2500°C for producing power inside an engine the temperature of the cylinder, in cylinder head, piston and valve, continues to rise when the engine runs. If these parts are not cooled by some means then they are likely to get damaged and even melted. The piston may cease inside the cylinder.

To prevent this, the temperature of the parts around the combustion, chamber is maintained at 200°C to 250°C. Too much of cooling will lower the thermal efficiency of the engine. Hence the purpose of cooling is to keep the engine at its most efficient operating temperature at all engine speeds and all driving conditions.

3.3.1 Water Cooling Systems

In water-cooling, water is used for cooling the engine by circulating it through water jackets around each combustion chamber cylinder, cylinder head, valve and valve sheet. By absorbing heat, water will become hot. When it is again passed through radiator, it will be cooled by air blast due to forward motion of the vehicle as well as of this engine to absorb heat.

There are two systems of water-cooling:

1. Thermosyphon system.
2. Pump circulation system.

1. Thermosyphon System

The principle of hot water going up and cold Water coming down due to difference in density is used here. There is no pump to circulate water. The light hot water from the engine goes to the top of the radiator by itself and gets cooled by the surrounding air and hence goes down to bottom of radiator and again goes to engine cylinder as shown in figure.

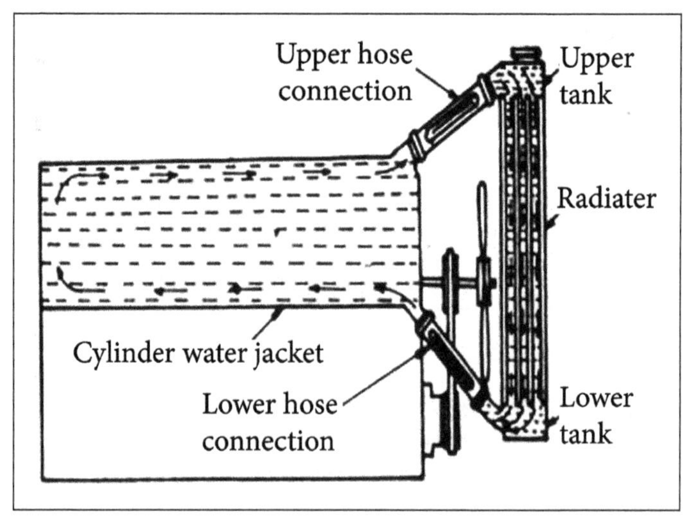

Thermosyphon system of cooling.

It is simple, cheap but cooling is slow. Water should be maintained to correct level at all time.

2. Pump Circulation System

To make the thermosyphon system more effective and improve water circulation, a water pump is introduced as shown in figure which is driven by a V-belt from a pulley on the engine crank shaft. This is called pump circulation system.

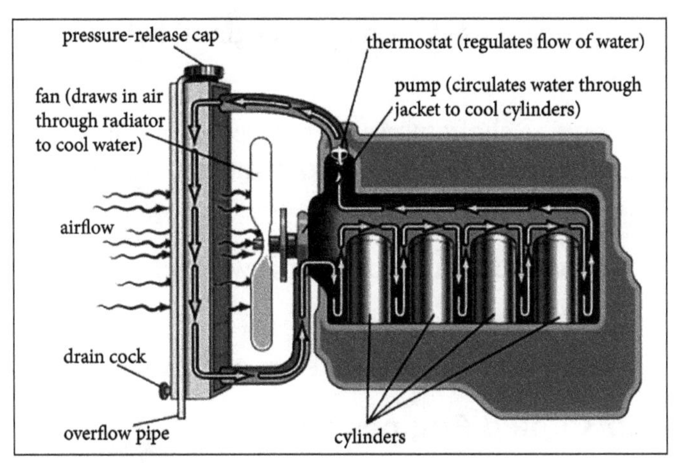

Water-cooling system for 4-cylinder engine.

The water-cooling arrangement for a 4 cylinder engine is shown in figure. When the hot water in engine passes through radiator tubes from upper tank to lower tank, it is exposed to large amount of airflow and gets sufficiently cooled.

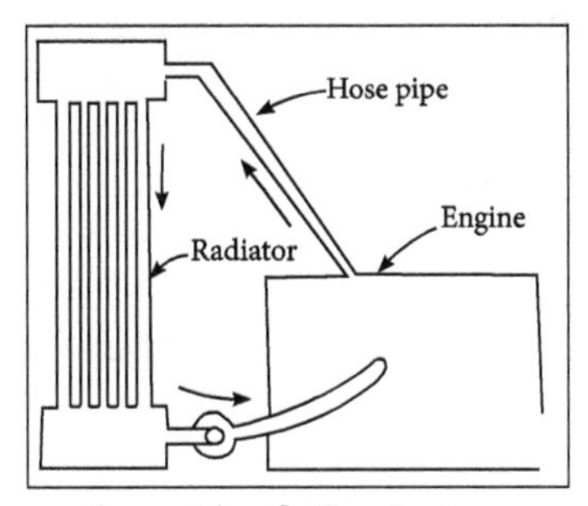

Pump Circulation System.

Then it is pumped to cylinder jackets by the water pump. The automatic thermostatic valve is used to regulate the circulation of water so that very cold water will become hot in short time to improve efficiency of the engine.

3.3.2 Air Cooling

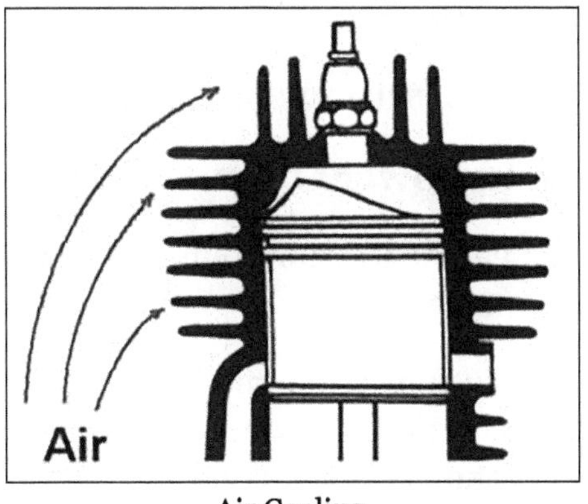

Air Cooling.

1. Air cooling is seldom used for marine engines, but may be used on auxiliary engines.
2. A fan on the engine shaft, blows air past 'cooling fins' on the engine castings. Metal cowling concentrates the air flow where it is most effective for removing heat.
3. Maintenance is reduced in keeping air cowls and cooling fins free of dust and rubbish.
4. Air cooled engines should be operated in open, well ventilated areas.

Advantages of Air Cooled Engines

1. Design of air-cooled engine is simple.
2. It is lighter in weight than water-cooled engines due to the absence of water jackets, radiator, circulating pump and the weight of the cooling water.
3. No risk of damage from frost, such as cracking of cylinder jackets or radiator water tubes.
4. It needs less care and maintenance.
5. This system of cooling is particularly advantageous where there are extreme climatic conditions in the arctic or where there is scarcity of water as in deserts.
6. It is cheaper to manufacture.

Disadvantages of Air Cooled Engines

1. Cooling is not uniform.
2. Relatively large amount of power is used to drive the cooling fan.
3. Cooling fins under certain conditions may vibrate and amplify the noise level.
4. Engines give low power output.
5. Engines are subjected to high working temperature.

3.3.3 Effect of Cooling on Power Output and Efficiency

1. Power output of an engine is proportional to volumetric efficiency provided the combustion is complete.
2. The volumetric efficiency of an engine is affected by many variables such as compression ratio, valve timing, induction and port design, mixture strength, latent heat of evaporation of the fuel, heating of the induced charge, cylinder temperature and the atmospheric conditions.

At higher speeds on account of internal effects the charge inhaled decreases and consequently the volumetric efficiency decreases. At lower speeds volumetric efficiency is nearly constant, at high speeds it falls rapidly.

At altitude the pressure and temperature both decrease. The effect of both is to reduce the density and consequently mass inhaled reduces. Thus higher the altitude lower will be the volumetric efficiency.

With normal aspiration the volumetric efficiency is seldom above 80% and to improve on this figure, supercharging is done. Air is forced into the cylinder by a blower or fan which is driven by the engine,

$$\text{Volumetric efficiency} = \frac{\text{Unit air charge}}{\text{Mass of air that would fill the displacement of one cylinder at inlet temperature and pressure.}}$$

Performance calculation

The power developed by an engine at the output shaft is called the brake power.

$$B_p = \frac{2\Pi NT}{60} \text{ in W}$$

$$N = \text{Speed in rpm}$$

$$T = \text{Torque in N} - \text{m}$$

Problems

1. During the trial (60 minutes) on a single cylinder oil engine having cylinder diameter 300 mm, stroke 450 mm and working on the four stroke cycle, the following observations were made:

Total fuel used: 9.6 liters

C.V. of the fuel: 45000 kJ/kg

Total No. of Revolutions: 12624

Gross IMEP: 7.24 bar

Pumping TMEP: 0.34 bar

Net load on the brake: 3150N

Diameter of the brake wheel drum: 1.78m

Diameter of the rope: 40 mm

Cooling water circulated: 545 liters

Cooling water temperature rise: 25°C

Specific gravity of oil: 0.8

Let us determine the indicated power, brake power and mechanical efficiency.

Solution:

Given:

$$k = 1; D = 1.78 \text{ m}; W = 3150 \text{ N}; D_c = 0.3\text{m}; L = 0.45\text{m}$$

Mean effective pressure,

$$P_m = P_{mg} - P_{mp}$$

$$= 7.24 - 0.34 = 6.9 \text{ bar}$$

$$P_m = 6.9 \text{ bar}$$

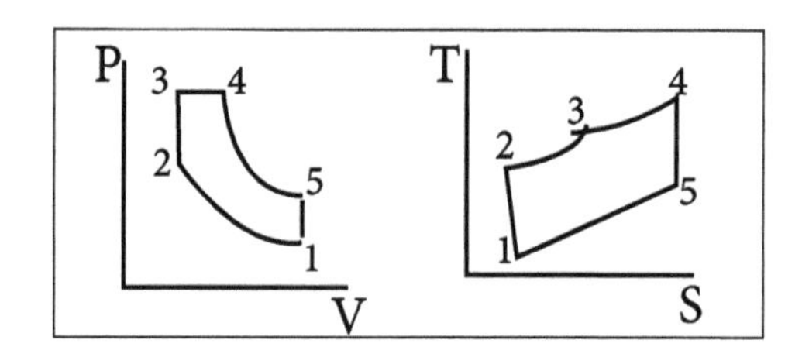

From process 1-2,

$$\frac{T1}{T2} = \frac{1}{(r)^{y-1}} = \frac{1}{(12)^{1.4-1.0}} = \frac{1}{(12)^{0.4}}$$

$$\Rightarrow \frac{T_1}{T_2} = (12)^{0.4}$$

$$T_2 = (12)^{0.4} \times 300 = 810.576 k$$

$$T_2 = 810.576 k$$

$$\frac{P_2}{P_1} = (r)^y = (12)^{1.4}$$

$$P_2 = (12)^{1.4} \times 1 \times 10^5 = 32.42 \times 10^5 \text{ N/m}^2$$

$$P_2 = 32.42 \times 10^5 \text{ N/m}^2$$

Pressure Ratio,

$$r_p = \frac{P_3}{P_2} = \frac{70 \times 10^5}{32.42 \times 10^5} = 2.159$$

$$r_p = 2.159$$

Efficiency of cycle,

$$\eta_{dual} = 1 - \frac{t}{12^{0.4}\left[\frac{2.159 \times 1.33^{1.4} - 1}{(2.159-1)+(1.4 \times 2.139 \times 0.33)}\right]}$$

$$\eta_{dual} = 61.97\%$$

Stroke volume,

$$V_s = \frac{\pi}{4} \times d^2 \times L$$

$$= \frac{\pi}{4} \times 0.25^2 \times 0.3$$

$$V_s = 0.0147\,m^3$$

w.k.t,

$$\gamma = \frac{V_s + V_c}{V_c}$$

$$V_c = \frac{V_s}{\gamma - 1} = \frac{0.0147}{12-1}$$

2 - 3 is constant volume process.

$$V_c = V_2, = 1.338 \times 10^{-3} m^3$$

$$V_2 = V_3 = 1.338 \times 10^{-3} m^3$$

$$\frac{P_2 V_2}{T_2} = \frac{P_3 V_3}{T_3}$$

$$T_3 = \frac{P_3}{P_2} \times T_2 = rP \times T_2$$

$$T_3 = 1750.03k$$

We know that,

$$\rho = \frac{V_4}{V_3} \Rightarrow V_4 = e \times V_3 = 1.33 \times 1.328 \times 10^{-3}$$

$$V_4 = 1.779 \times 10^{-3} m^3$$

From process 3 - 4; (P = C),

$$\frac{P/3V_3}{T_3} = \frac{P/4V_4}{T_4}$$

$$T_4 = \frac{V_4}{V_3} \times T_3 = \rho \times T_3 = 1.33 \times 1750.03$$

$$T_4 = 2327.54 \text{ K}$$

$$\text{Speed } N = \frac{12624}{60} = 210.4 \text{ rpm}$$

For four stroke engine $n = \frac{N}{2}$

Indicated Power:

$$I.P = \frac{P_m LA \times n \times k}{60}$$

$$= \frac{6.9 \times 10^5 \times 0.45 \times \frac{\pi}{4} \times 0.3^2 \times \frac{210.4}{2} \times 1}{60}$$

$$I.P = 38.428 \text{ Kw}$$

Brake Power:

$$B.P = \frac{(W - S)\pi D_6 N}{60 \times 1000}$$

$$= \frac{3150 \times \pi \times 1.78 \times 210.4}{60 \times 1000}$$

$$B.P = 61.77 \text{ kW}$$

2. In a constant speed compression ignition engine operating on four- stroke cycle and fitted with band brake, the following observations were taken:

Brake wheel diameter 60 cm,

Band thickness 5 mm,

Speed 450 rpm,

Load on band210 N,

Spring balance reading 30 N,

Area of indicator diagram 4.15 cm^3,

Length of indicator diagram 6.25 cm^3,

Spring No. 11, i.e. 11 bar/cm,

Bore=10 cm,

Stroke=15 cm,

Specific fuel compression 0.3 kg/kW-hr,

Heating value of fuel 41800 kJ/kg.

Let us determine the brake power, indicated power and mechanical efficiency:

Solution:

Given:

$$D = 60 \text{ cm} = 0.6 \text{ m}$$

$$N = 450 \text{ rpm} = 7.5 \text{ rps}$$

$$W_1 = 210 \text{ N} = 0.21 \text{ kN}$$

$$W_2 = 30 \text{ N}$$

$$A = 41.5 \text{ cm}^2$$

$$L = 6.25 \text{ cm}$$

$S = 11 \text{ bar} / \text{cm}$

$d = 10 \text{ cm} = 0.1 \text{ m}$

$I = 15 \text{ cm} = 0.15 \text{ m}$

$m_f = 0.38 \text{ kg} / \text{ kw hr}$

$C_v = 41800 \text{ kJ} / \text{kg}$

To find:

1. BP
2. IP
3. η_{mech}

Brake Power BP $= 2\pi NWR$

Net load on brake drum $W = W_1 - W_2$

$= 210 - 30$

$= 180 \text{ N}$

0.18 kN

Effective brake radius, $R = \frac{D+d}{2} = \frac{0.6+(5\times10^{-3})}{2}$

$R = 0.3025 \text{ m}$

Break Power BP $= 2\pi N \, WR$

$= 2\pi\times7.5\times0.18\times0.3025$

$BP = 2.56589 \text{ kW}$

Indicated Power $I_p = P_m l\times a\times n\times k$

Indicated mean effective pressure P_m

$$P_m = \frac{AS}{L}$$

$$= \frac{4.15\times11}{6.25}$$

$$= 7.034\,\text{bar}$$

$$P_m = 730.4\,\text{kN}/\text{m}^2$$

$$I_p = P_m l \times a \times n \times k$$

$$a = \frac{\pi}{4} d^2 = \frac{\pi}{4} \times 0.1^2$$

$$a = 7.853 \times 10^{-3}\,\text{m}^2$$

$$I_p = P_m l \times a \times n \times k$$

$$= 730.4 \times 0.15 \times 7.853 \times 10^3 \times 4 \times 1$$

$$I_p = 13.767\,\text{kW}$$

Mechanical Efficiency:

$$\eta_{mech} = \frac{BP}{IP} = \frac{2.56589}{13.767}$$

3.4 Properties of Lubricant

Viscosity: It is the ability of the oil to resist internal deformation due to mechanical stresses and hence it is a measure of the ability of the oil film to carry a load. More viscous oil can carry a greater load, but it will offer greater friction to sliding movement of the one bearing surface over the other.

Viscosity varies with the temperature and hence if a surface to be lubricated is normally at high temperature it should be supplied with oil of a higher viscosity.

Flash Point: It is defined as the lowest temperature at which the lubricating oil will flash when a small flame is passed across its surface. The flash point of the oil should be sufficiently high so as to avoid flashing of oil vapours at the temperatures occurring in common use. High flash point oils are needed in air compressors.

Fire Point: It is the lowest temperature at which the oil burns continuously. The fire point also must be high in lubricating oil, so that oil does not burn in service.

Cloud Point:When subject to low temperatures the oil changes from liquid state to a plastic or solid state. In some cases the oil starts solidifying which makes it to appear cloudy. The temperature at which this takes place is called the cloud point.

Pour Point: Pour point is the lowest temperature at which the lubricating oil will pour. It is an indication of its ability to move at low temperatures. This property must be considered because of its effect on starting an engine in cold weather and on free circulation of oil through exterior feed pipes when pressure is not applied.

Oiliness: This is the property which enables oil to spread over and adhere to the surface of the bearing. It is most important in boundary lubrication.

Corrosion: A lubricant should not corrode the working parts and it must retain its properties even in the presence of foreign matter and additives.

Emulsification: Lubricating oil, when mixed with water is emulsified and loses its lubricating property. The emulsification number is an index of the tendency of an oil to emulsify with water.

Physical Stability: Lubricating oil must be stable physically at the lowest and highest temperatures between which the oil is to be used. At the lowest temperature there should not be any separation of solids and at the highest temperature it should not vapourise beyond a certain limit.

Chemical Stability: Lubricating oil should also be stable chemically. There should not be any tendency for oxide formation.

Rationalization Number: Oil may contain certain impurities that are not removed during refining. The neutralization number test is a simple procedure to determine acidity or alkalinity of an oil. It is the weight in milligrams of potassium hydroxide required to neutralize the acid content of one gram of oil.

Adhesiveness: It is the property of lubricating oil due to which the oil particles stick with the metal surfaces.

Film Strength: It is the property of lubricating oil due to which the oil retains a thin film between the two surfaces even at high speed and load. The film does not break and the two surfaces do not come in direct contact.

Adhesiveness and film strength cause the lubricant to enter the metal pores and cling to the surfaces of the bearings and journals keeping them wet when the journals are at rest and presenting metal to metal contact until the film of lubricant is built up.

Specific Gravity: It is a measure of density of oil. It is an indication regarding the grade of lubricant by comparing one lubricant with other. It is determined by a hydrometer which floats in the oil and the gravity is read on the scale of the hydrometer at the surface of the oil.

3.4.1 Different Types of Lubricating System

1. Wet sump lubrication system.
    i. Splash System.
    ii. Semi-pressure system.
    iii. Full pressure system.
2. Dry sump lubrication system.
3. Mist lubrication system.

1. Wet Sump Lubrication System

These systems employ a large capacity oil sump at the base of crank chamber, from which the oil is drawn by a low pressure oil pump and delivered to various parts. Oil then gradually returns back to the sump after serving the purpose.

(i) Splash System

This system is used on some small four-stroke stationary engines. In this case the caps on the big ends bearings of connecting rods are provided with scoops which, when the connecting rod is in the lowest position, just dip into oil troughs and thus direct the oil through holes in the caps to the big end bearings.

Due to splash of oil, oil reaches the lower portion of the cylinder walls, crankshaft and other parts requiring lubrication. Surplus oil eventually flows back to the oil sump. Oil level in the trough is maintained by means of a oil pump which takes oil from sump, through a filter. Splash system is suitable for low and medium speed engines having moderate bearing load pressures. For high performance engines, which normally operate at high bearing pressures and rubbing speeds this system does not serve the purpose.

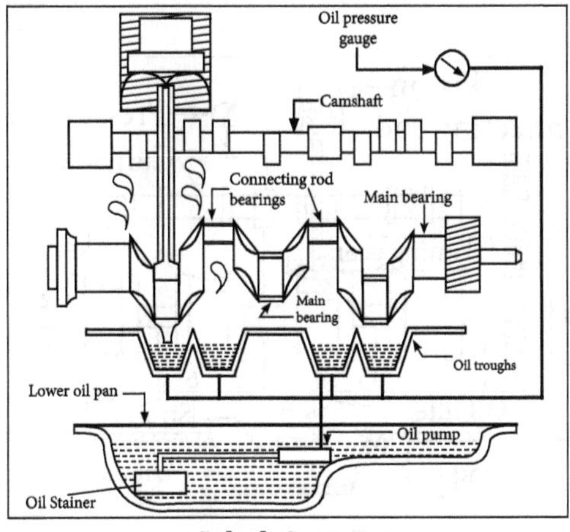

Splash System.

(ii) Semi-Pressure System

This method is a combination of splash and pressure systems. It incorporates the advantages of both. In this case main supply of oil is located in the base of crank chamber. Oil is drawn from the lower portion of the sump through a filter and is delivered by means of a gear pump at pressure of about 1 bar to the main bearings. The big end bearings are lubricated by means of a spray through nozzles. Thus oil also lubricates the cams, crankshaft bearings, cylinder walls and timing gears. An oil pressure gauge is provided to indicate satisfactory oil supply.

The system is less costly to install as compared to pressure system. It enables higher bearing loads and engine speeds to be employed as compared to splash system.

(iii) Full Pressure System

In this system, oil from oil sump is pumped under pressure to the various parts requiring lubrication, shown in figure (a). The oil is drawn from the sump through filter and pumped by means of a gear pump. Oil is delivered by the pressure pump at pressure ranging from 1.5 to 4 bar. The oil under pressure is supplied to main bearings of crankshaft and camshaft.

Holes drilled through the main crankshafts, bearing journals, communicate oil to the big end bearing and also small end bearings through hole drilled in connecting rods. A pressure gauge is provided to confirm the circulation of oil to the various parts. A pressure regulating valve is also provided on the delivery side of this pump to prevent excessive pressure.

This system finds favor from most of the engine manufacturers as it allows high bearing pressure and rubbing speeds.

The general arrangement of wet sump lubrication system is shown in figure (b). In this case oil is always contained in the sump which is drawn by the pump through a strainer.

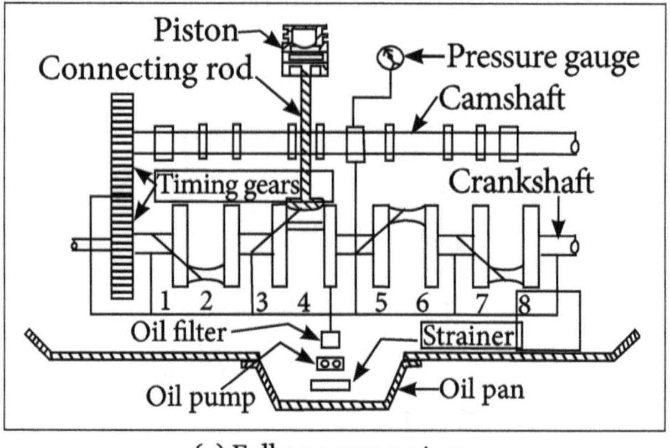

(a) Full pressure system.

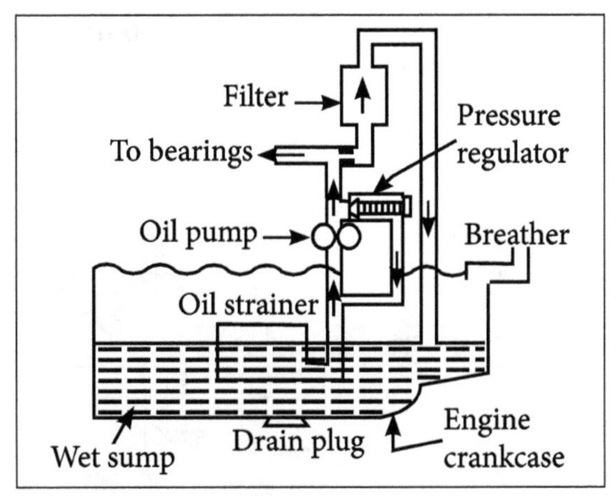

(b) Wet Sump Lubrication System.

2. Dry Sump (Lubrication) System

An engine lubrication system is the one in which the lubricating oil is carried in an external tank and not internally in a sump. The sump is kept relatively free from oil by scavenging pumps, which return the oil to the tank after cooling. The opposite of a wet sump system. The pumping capacity of scavenge pumps is higher than that of the engine-driven pumps supplying oil to the system.

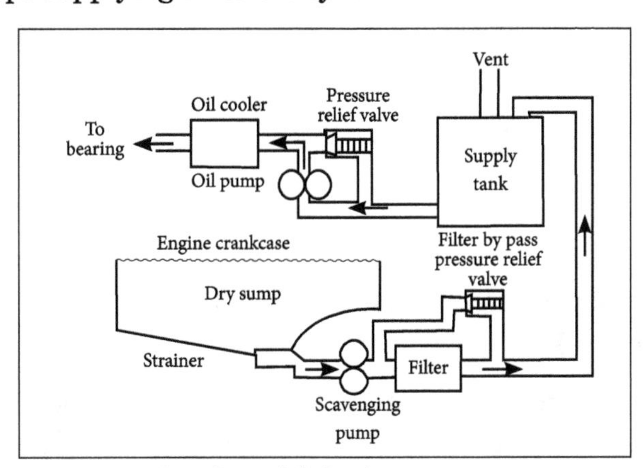

Dry Sump lubrication system.

1. This system is used for two stroke cycle engines. Most of these engines are crank charged, i.e., they employ crankcase compression and thus, are not suitable for crankcase lubrication.

2. These engines are lubricated by adding 2 to 3 per cent lubricating oil in the fuel tank. The oil and fuel mixture is induced through the carburettor. The gasoline is vaporized; and the oil in the form of mist, goes via crankcase into the cylinder. The oil which impinges on the crankcase walls lubricates the main and connecting rod bearings and rest of the oil which passes on the cylinder during charging and scavenging periods, lubricates the piston, piston rings and the cylinder.

3. For a good performance, F/A ratio used is also important. A F/A ratio of 40 to 50:1 is optimum. Higher ratios increase the rate of wear and lower ratios result in spark plug fouling.

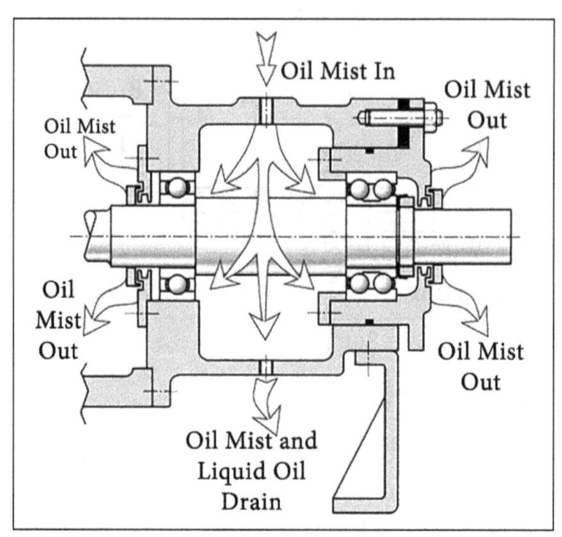

Oil Mist Introduced at Midpoint of Bearing Housing.

Advantages

1. System is simple.
2. Low cost (because no oil pump, filter etc. are required).

Disadvantages

1. Some lubrication oil will burn and cause heavy exhaust emissions and deposits on piston crown, ring grooves and exhaust port and thus hamper the good performance of the engine.
2. Since the lubricating oil comes in contact with acidic vapours produced during the combustion process, it rapidly loses its anti-corrosion properties resulting in corrosion damage of bearing.
3. The oil and fuel must be thoroughly mixed for effective lubrication. This requires either separate mixing prior to use or use of some additive to give the oil good mixing characteristics.
4. Owing to higher exhaust temperature and less efficient scavenging the crankcase oil is diluted. In addition some lubricating oil burns in combustion chamber. This results in 5 to 15 per cent higher lubricant consumption for two stroke engine of similar size.
5. As there is no control over the lubricating oil, once introduced with fuel, most of the two stroke engines are over-oiled most of the time.

3.5 Engine Emission and Control: Mechanism of Pollutant Formation and its Harmful Effects

The major pollutants emitted from the exhaust due to incomplete combustion are:

1. Carbon monoxide (CO).
2. Hydrocarbons (HC).
3. Oxides of nitrogen (NO).

Mechanism of Pollutant Formation

Petrol is a complex blend of hydrocarbons such as paraffins, cycloparaffins, olefins and aromatics, which contain mainly carbon and hydrogen atoms. In an automobile engine these hydrocarbon compounds are burnt with air to produce energy for propelling the vehicle. Burning is the chemical process of oxidization, in which the hydrocarbons are combined with the oxygen in air. If sufficient oxygen is available, the hydrocarbon is completely oxidized so that all of the carbon atoms combine with oxygen to produce CO_2 (carbon dioxide) and all of the hydrogen atoms combine with oxygen to produce H_2O (water) and both of these substances are harmless. $HC + O_2 + N_2 \longrightarrow CO_2 + H_2O + N_2$ For stoichiometric (chemically perfect) combustion an air-fuel ratio of about 14.7:1 is required. This means for complete burning of 1 gm of petrol, consisting mainly of heptane (H_7H_{16}) and hexane (C_6H1_4), about 14.7 gm of air are required (a volume of about 12.2 litres).

In practice, air-fuel mixtures may be either fuel-rich or fuel-lean. Therefore, the ratio of the actual air-fuel ratio to the stoichiometric air-fuel ratio is considered as a useful parameter for describing mixture strength. This parameter, termed as relative air-fuel ratio or excess air factor, is denoted by the symbol X (the Greek letter lambda) and is defined as,

λ = (actual air fuel ratio) (stoichiometric air-fuel ratio),

Where,

$\lambda>1$,-Fuel lean mixtures,

$\lambda=1$,-Fuel lean mixtures,

And $\lambda<1$-Fuel rich mixtures.

A complete oxidization rarely takes place in a spark-ignition engine even with a stoichiometric mixture, because the hydrocarbons are forced to burn in a short period of time with a fixed volume of air. Atmospheric nitrogen, which is normally chemically inert, starts to react with oxygen at high temperature and pressure prevailing in the combustion chamber.

The nitrogen thus consumes oxygen, which otherwise react with the hydrocarbons, leading to their incomplete combustion. This resulted in the formation of undesirable combustion products, specifically the pollutants CO (carbon monoxide), HC (unburned hydrocarbon) and various oxides of nitrogen (NO, NO2 and N2O, collectively termed NOx), i.e.

$$HC + O_2 + N_2 \longrightarrow CO_2 + CO + H_2O + HC + N_2 + NO_X$$

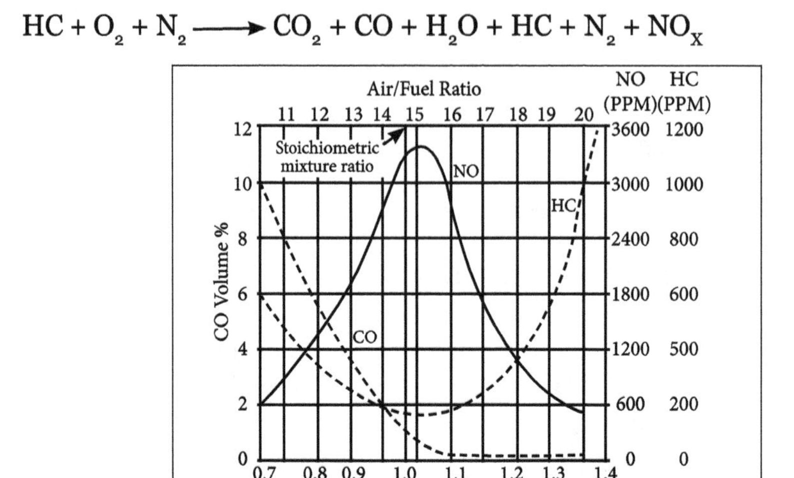

Pollutant emission as a function of relative air-fuel ratio.

Even with a stoichiometric mixture (X = 1) pollutant gases (HC, CO and NOx) are produced, which constitute up to 2 percent of the exhaust gas. The relative quantities of the pollutants produced by the engine depend principally upon the relative air-fuel mixture ratio and vary in the manner illustrated in figure. Maximum combustion temperature occurs at slightly fuel lean mixture of about X = 1.05 and decreases rapidly to either side.

Since NOx emission are temperature dependent they tend to follow this characteristic and so decrease rapidly as X increases. On the other hand HC emissions increase away from stoichiometric.

On the rich side (X < 1) this arises as an outcome of incomplete combustion due to an inadequate oxygen supply. On the lean side (X > 1) slow or incomplete burning causes the exhaust valve to open before combustion is entirely completed, allowing some unburned HC into the atmosphere. With very lean mixtures, misfire may occur so that HC emissions rise sharply.

The harmful effects of detonation are as follows:

1. Noise and Roughness

The knocking produces a loud pulsating noise and pressure waves. These waves vibrate back and forth across the cylinder. The presence of vibratory motion causes crankshaft vibrations and the engine runs rough.

2. Mechanical Damage

i. Maximum pressure waves generated during knocking can increase rate of wear of parts of combustion chamber. Severe erosion of piston crown, cylinder head and pitting of inlet and outlet valves may result in complete wreckage of the engine.

ii. Detonation is very dangerous in engines having high noise level. In small engines the knocking noise is easily detected and the corrective measures can be taken but in aero engines it is difficult to detect knocking noise and hence corrective measures cannot be taken. Hence severe detonation may persist for a long time which may ultimately result in complete wreckage of the piston.

3. Carbon Deposits

Detonation results in increased carbon deposits.

4. Increase in Heat Transfer

Knocking is accompanied by an increase in the rate of heat transfer to the combustion chamber walls:

i. The increase in heat transfer is due to two reasons.

ii. The minor reason is that the high temperature in a detonating engine is about 150°C higher than in a non-detonating engine, due to rapid completion of combustion.

iii. The major reason for increased heat transfer is the scouring away of protective layer of inactive stagnant gas on the cylinder walls due to pressure waves. The inactive layer of gas normally reduces the heat transfer by protecting the combustion and piston crown from direct contact with flame.

5. Decrease in Power Output and Efficiency

Due to increase in the rate of heat transfer the power output as well as efficiency of a detonating engine decreases.

6. Pre-Ignition

Increase in the rate of heat transfer to the walls has yet another effect. It may cause local overheating, especially of the sparking plug, which may reach a temperature high enough to ignite the charge before the passage of spark, thus causing pre-ignition. An engine detonating for a long period would most probably lead to pre-ignition and this is the real danger of detonation.

3.5.1 Methods of Measuring Pollutants and Control of Engine Emission

3-Way Catalytic Converter

Most cars today are equipped with a 3-way catalytic converter. There, 3-way refers to the three emissions it helps to reduce, carbon monoxide, hydrocarbons or volatile organic compounds (VOC's) and NOx molecules.

The 3-way converts use two different types of catalysts, a reduction catalyst and an oxidization catalyst. Each types consist of a base structure coated with a catalyst such as platinum, rhodium and palladium. The scheme is to create a structure that exposes the maximum surface area of the catalyst of the exhaust flow, while minimizing the amount of catalyst required.

3-way converters use two catalyst processes:

1. Reduction
2. Oxidation

A sophisticated oxygen storage or engine control system to convert three harmful gasses -HC, CO and oxides of nitrogen (NO_3). This is not an easy task, the catalyst chemistry required to clean up NO_2 is most effective with a rich air or fuel mix, whereas HC and CO reduction are most effective with a lean air fuel bias. To operate properly, a three - way converter first must convert NOx (with a rich air and fuel bias), then HC and CO (with a lean bias).

Reduction Catalyst

The reduction catalyst is the first stage of the catalytic converter. It uses platinum and rhodium to help reduce the Nx emissions. When an NO or NO3 molecule contacts the catalyst, the catalyst rips the nitrogen atom out of the molecule and holds on to it, freeing the oxygen in the form of O_2. The nitrogen atoms bond with other nitrogen atoms that are also stuck to the catalyst forming N_2.

$$2NO \Rightarrow N_2 + O_2$$

$$2NO_2 \Rightarrow N_2 + 2O_2$$

Oxidation Catalyst

The oxidation catalyst is the second stage of the catalytic converter. It reduces the unburnt hydrocarbons and carbon monoxide by burning (oxidizing) them over a platinum and palladium catalyst. This catalyst aids the reaction of the CO and hydrocarbons with the remaining oxygen in the exhaust gas. An example:

$$2CO + O_2 \Rightarrow 2CO_2$$

Vehicles with catalyst converter must use unleaded petrol. Lead in petrol contaminate the catalyst and makes it ineffective. For the catalytic converter to most effective, the air fuel mixture must have stoichiometric ratio of 14:7:1.

To achieve the described air fuel ratio at all operating conditions, a feedback system is used. It determines the correct air fuel ratio of the intake charge by measuring the amount of oxygen remaining in the exhaust.

The diesel engine catalytic converter is a pure oxidation catalytic converter. It oxidizes HC and CO into water and CO_2. It cannot reduce NO_2.

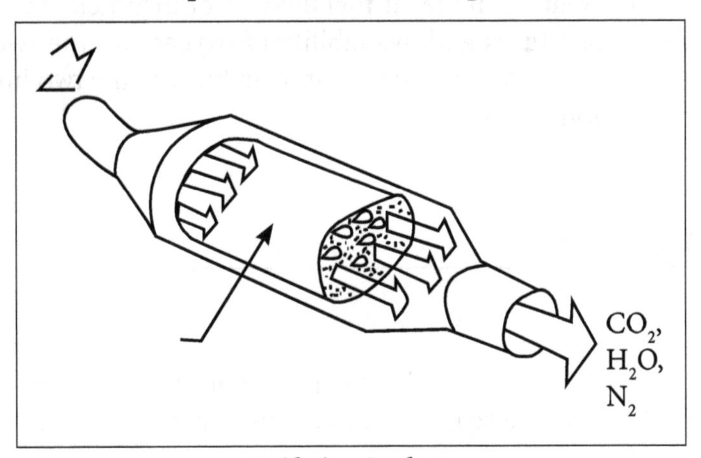

Oxidation Catalyst.

Emission Norms

Pollutants

There are three main sources of air pollution due to petrol engine:

1. Evaporator losses through carburetor.
2. Crank case blow by.
3. Exhaust emission through the pipe.

1. Evaporator Emission

Evaporator emission takes place from the fuel supply system. The main reason of hydrocarbon evaporation is maximum temperature. Fuel volatility, location of tank, layer of fuel line, mode of operation also affect the evaporation, About 30% of the total hydrocarbon emission is occurring from the fuel tank, fuel line and carburetor.

2. Crankcase Blow By

Crank case blow by means the leakage part of the piston and piston rings from the cylinder to the crank case. In blow by gases, there are about 85% or low HC and rest 15% of the current gases. It is about 20% of the total HC emission from the engine and may be 30% of the rings are worn out. The blow by gases are controlled by the crankcase ventilation system.

3. Exhaust Emission

The exhaust emission contains HC, CO and NO_2 (oxide of nitrogen). It occurs due to incomplete combustion. The emission of HC is closely related to many designing and operating factors like induction system, combustion chamber design, air fuel ratio speed, load and mode of operation. Lean mixture given lower HC emission.

C_0 occurs due to insufficient air in the air fuel mixture or insufficient time for complete combustion. High temperatures and availability of oxygen are the two main reasons for the formation of NO_2. The spark advance and air-fuel are the two important factors which affect the formation of No_2.

3.6 Gas Turbines

Introduction

A simple gas turbine unit consists of three components, viz., a compressor, a heat addition device and a turbine. These three components can be arranged either in an open or a closed form.

Accordingly, a gas turbine cycle can be classified into two categories:

1. Open cycle
2. Closed cycle

The two, open-cycle arrangements are much more common. In this arrangement fresh atmospheric air is drawn into the system continuously and energy is added by combustion of fuel in the working fluid itself. The products of combustion are expanded through the turbine and exhausted into the atmosphere.

In the closed-cycle, the same working fluid, be it air or some other gas, is repeatedly circulated through the system. It may be noted that in this type of plant whether the working fluid is air or some other gas, fuel cannot be burnt directly in the working fluid and the necessary energy must be added in a heater or gas boiler.

Open Cycle Gas Turbine Engine

The open gas turbine cycle described above can be modeled as a closed cycle, by utilizing the air-standard assumptions. Here the compression and expansion processes remain the same but the combustion process is replaced by a constant pressure heat addition process from an external source and the exhaust process is replaced by a constant pressure heat rejection process to the ambient air.

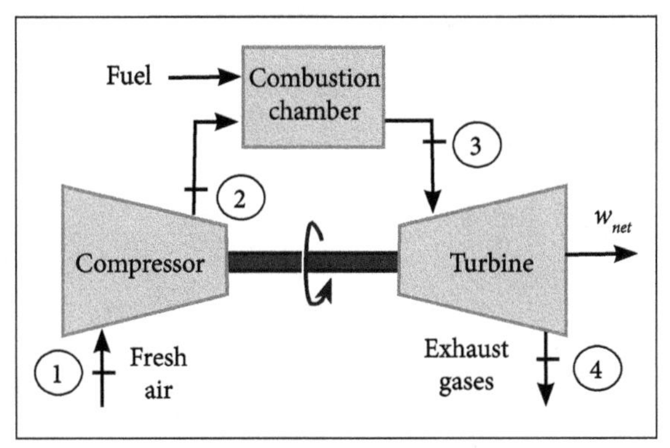

Open cycle gas turbine.

The ideal cycle that the working fluid undergoes in this closed loop is the Brayton cycle, which is made up of four internally reversible processes:

1-2 Isentropic compression (in a compressor).

2-3 Constant-pressure heat addition.

3-4 Isentropic expansion (in a turbine).

4-1 Constant-pressure heat rejection.

Close Cycle Gas Turbine Engine

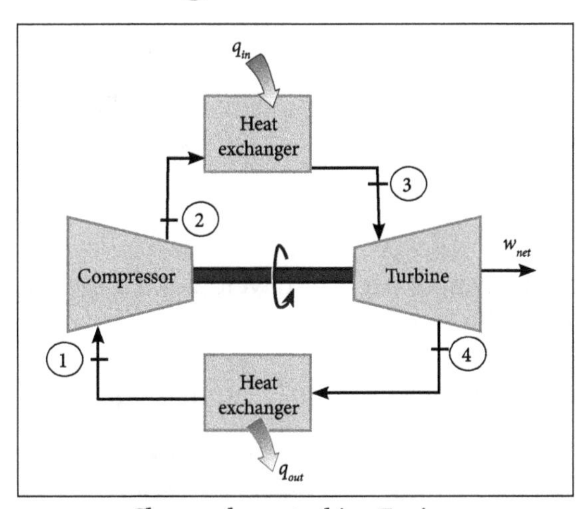

Close cycle gas turbine Engine.

Because of fixed amount of gas, the closed cycle gas turbine is not an internal combustion engine. In the closed cycle system, combustion cannot be sustained and the normal combustor is replaced with a second heat exchanger to heat the compressed air before it enters into the turbine.

The heat is supplied by an external source such as a nuclear reactor, the fluidized bed of a coal combustion process or some other heat source. Closed cycle systems using gas turbines have been proposed for missions to Mars and other long term space applications.

Advantages of Closed Cycle

1. Higher thermal efficiency.
2. Reduced size.
3. No contamination.
4. Improved heat transmission.
5. Improved part load.
6. Lesser fluid friction.
7. No loss of working medium.
8. Large output.
9. Less fuel.

Disadvantages of Closed Cycle

1. Quality.
2. Large amount of cooling water is needed. This limits its use of stationary installation or marine use.
3. Dependent system.
4. The weight of the system pre kW developed is high comparatively, therefore not economical for moving vehicles.
5. Requires the use of a very large air heater.

We have thermal efficiency, $\eta_{th} = \frac{W_N}{Q_H} = \frac{Q_H - Q_L}{Q_H} = 1 - \frac{Q_L}{Q_H}$

Methods to Improve Thermal Efficiency

Heat added $Q_H = h_3 - h_2 = C_p(T_3 - T_2)$

Heat rejected $Q_L = h_4 - h_1 = C_p(T_4 - T_1)$

$$\therefore \eta_{th} = 1 - \frac{C_P(T_4 - T_1)}{C_P(T_3 - T_2)} = 1 - \frac{T_1\left[\frac{T_4}{T_1} - 1\right]}{T_2\left[\frac{T_3}{T_2} - 1\right]}$$

$$\text{Now, } \frac{T_2}{T_1} = \left(\frac{P_2}{P_1}\right)^{\frac{r-1}{r}} = R^{\frac{r-1}{r}} \;\&\; T_3/T_4 = \left(P_3/P_4\right)^{\frac{r-1}{r}} = \left(\frac{1}{R}\right)^{\frac{r-1}{r}}$$

But as $P_2 = P_3$ & $P_1 = P_4$, it follows that $\frac{T_2}{T_1} = \frac{T_3}{T_4}$ or $\frac{T_4}{T_1} = \frac{T_3}{T_1}$

$$\therefore \eta_{th} = 1 - \frac{T_1}{T_2} \quad \text{i.e., } \eta_{th} = 1 - \frac{1}{\left(\frac{T_2}{T_1}\right)} \quad \text{or} \quad \eta_{th} = 1 - \frac{1}{R^{\frac{r-1}{r}}}$$

3.7 Air Craft Propulsion: Analysis of Turbo Jet

Thrust Augmentation

To achieve better take-off performance, higher rates of climb and increased performance at altitude during combat maneuverer there has been a demand for increasing the thrust output of aircraft for short intervals of time. This is achieved by additional fuel in the tail pipe between the turbine exhaust and entrance section of the exhaust nozzle. This method of thrust increases the jet velocity, which is called Thrust Augmentation.

Construction of Turbo Jet Engine

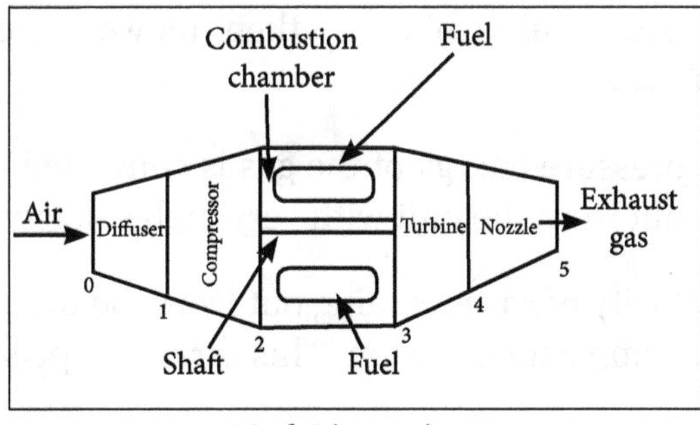

Turbo jet engine.

It consists of:

1. Diffuser.

2. Rotary compressor.
3. Combustion chamber.
4. Turbine.
5. Exhaust nozzle.

Principle and Working

The air-breathing engines described so far are simple in construction and they have not been used very extensively. The most common type of air-breathing engine is the Turbojet engine.

The function of the diffuser is to convert the kinetic energy of the entering air into pressure energy. The function of the nozzle is to convert the pressure energy of the combustion gases into kinetic energy. Now let us see the working:

1. Air from the atmosphere enters into turbojet engine. The air velocity gets reduced and its static pressure is increased by diffuser.
2. Then the air passes through the rotary compressor in which the air is further compressed.
3. Then the high pressure air flows into the combustion chamber where the fuel is injected by suitable injectors and the air-fuel mixture is burnt. Heat is supplied at constant pressure.
4. The highly heated products of combustion gases then enter the turbine and partially expanded.
5. The power produced by the turbine is just sufficient to drive the compressor, fuel pump and other auxiliaries.
6. The hot gases from the turbine are then allowed to expand in the exhaust nozzle section.
7. In the nozzle, pressure energy of the gas is converted into kinetic energy. So the gas comes out from the unit with very high velocity.
8. Due to high velocity of gases coming out from the unit, a reaction or thrust is produced in the opposite direction. This thrust propels the air craft.
9. Like ramjet engine, the turbojet engine is a continuous flow engine.
10. Because of turbine material limitations, only a limited amount of fuel can be burnt in the combustion chamber.

Thermodynamic Cycle

The turbojet engine works on the Brayton cycle. In the analysis of a turbo jet cycle, the following assumptions are made:

1. Specific heat is constant.
2. There is no pressure loss in the combustion chamber.
3. The power produced by the turbine is just sufficient to drive the compressor, fuel pump and other auxiliaries.

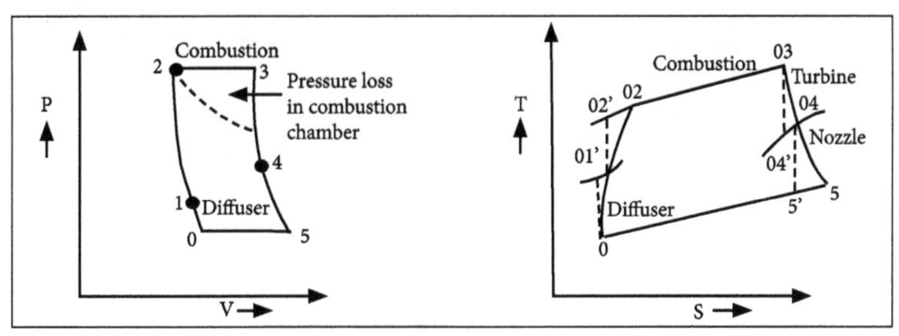

Turbo cycle.

Performance of a Turbo Jet Engine

Thrust specific fuel consumption (TSFC) vs. compressor pressure ratio (r) at two different Mach number is shown in the below figure:

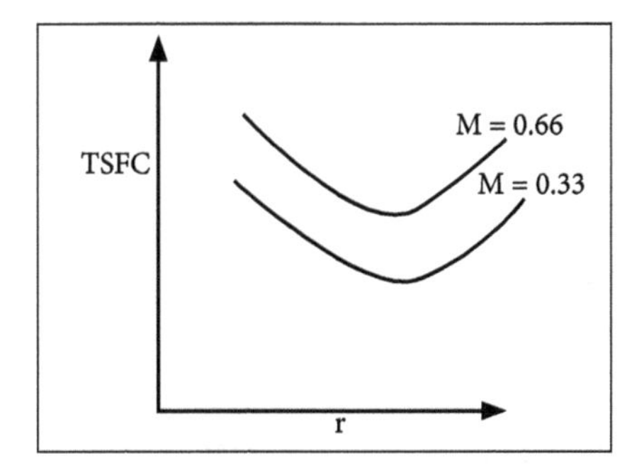

When the pressure ratio increases, the fuel consumption decreases up to certain value. After that further increase in pressure ratio leads to increase in fuel consumption.

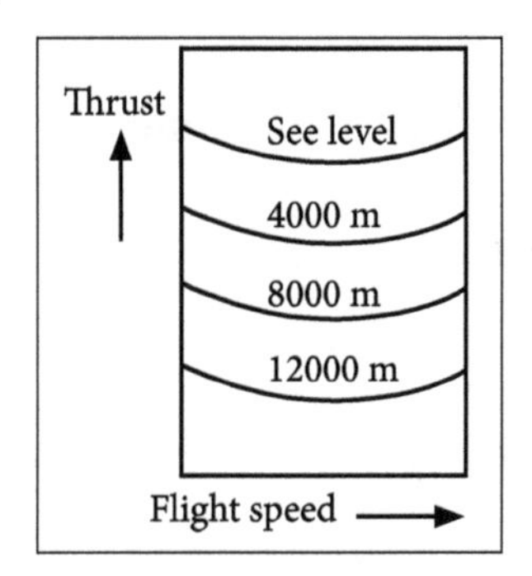

As the altitude increases, the thrust decreases due to decrease in density, pressure and temperature of air. But the rate of decrease of thrust is less than the rate of decrease of density with altitude.

Advantages

1. Construction is simple.
2. Less wear and tear.
3. Less maintenance cost.
4. It runs smoothly because continuous thrust is produced by continuous combustion of fuel.
5. The speed of a turbo jet is not limited by the propeller and it can attain higher flight speed than turbo propeller air crafts.
6. Low grade fuels like kerosene, paraffin, etc., can be used. This reduces the fuel cost.
7. Reheat is possible to increase the thrust.
8. Since turbo jet engine has a compressor, it can be operated under static conditions.

Disadvantages

1. It has low takeoff thrust and hence poor starting characteristics.
2. Fuel consumption is high.
3. Costly materials are used.
4. The fuel economy at low operational speed is extremely poor.
5. Sudden decrease of speed is difficult.
6. Propulsive efficiency and thrust are lower at lower speeds.

Applications

It is best suited for piloted air-crafts, military air crafts, etc.

Equations for Turbojet Engine

Jet Thrust

The control surface of a turbojet engine between section 1 and 2 is shown in the below figure.

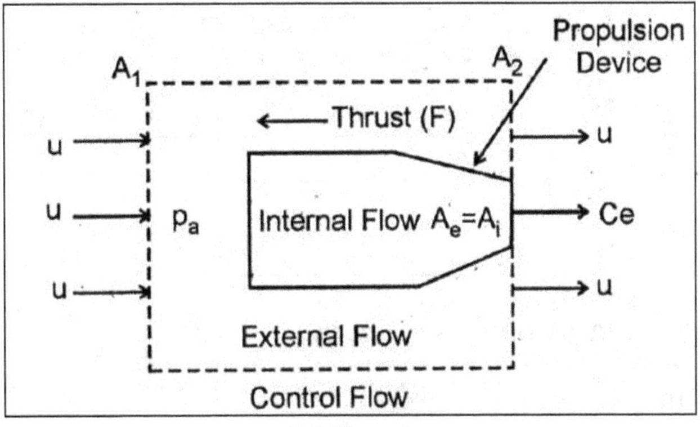

Jet Thrust.

Air from atmospheric enters the Turbojet engine at a pressure Pa and velocity u.

1. Net thrust of the engine,

$$(F)=\{\text{Momentum thrust}(F_{mom})\}+\{\text{Pressure thrust}(F_{pre})\}$$

2. Moment thrust $(F_{mom}) = (m_a + m_j)C_e - m_a \times u$

3. Pressure thrust $(F_{Pr}) = (P_e - P_a) \times A_e$

4. Net thrust

$$= (m_a + m_f - m_a - m_a X_u + (P_e - P_a) \times A_e P_a = P_e$$

Net thrust $F = (m_a + m_f)\, C_e - m_a \times u$

Where,

M_a – Mass of air (kg/s)

M – Mass of fuel (kg/s)

C_e – Exit velocity of gasses

C_e – Jet velocity

u – Flight speed

Propeller Thrust

The control surface of a turboprop engine between section 1 and 2 is shown in the below figure.

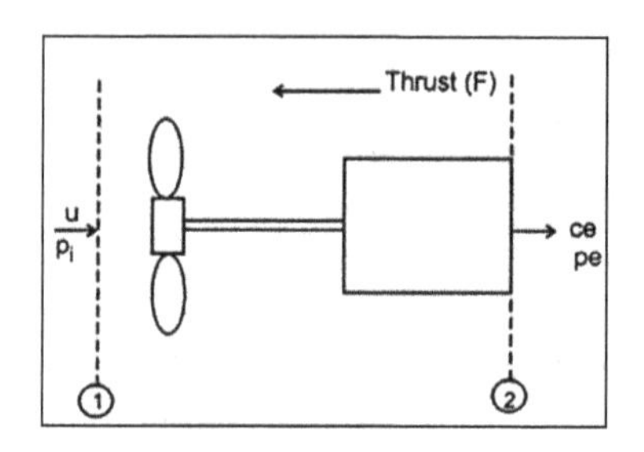

Net thrust considering mass of fuel is given by,

$$F = (m_a + m_f)\,C_e - m_a - m_a \times u$$

$$= m\,C_e - m_a \times u\,[\because m = m_a + m_f]$$

For complete expansion,

$$C_e = C_j$$

$$F = m\,c_j - m_a x\,u.$$

Net thrust without considering mass of fuel is given by,

$$F = m\,C_j - m_u$$

Or

$$F = \dot{m}_a\left(C_j - u\right)$$

Effective Speed Ratio (σ)

The ratio of flight speed to jet velocity is known as effective speed ratio (r),

$$\sigma = \frac{\text{First Speed}}{\text{Jet Velocity (or) Velocity of Exit gases}}$$

$$\sigma = \frac{u}{C_j}$$

$$\text{Thrust } F_1 = m_a\left[C_j - u\right]$$

$$= \dot{m}_a \times C_j\left[1 - \frac{u}{C_j}\right]$$

$$F = \dot{m}_a \times C_j\left[1 - \sigma\right]$$

Propulsive Efficiency

It is defined as the ratio of propulsion power (or) thrust power to the power output of the engine.

$$\eta_p = \frac{\text{Propulsive Power (or) Thrust Power}}{\text{Power output of the engine}}$$

$$\text{Thrust Power} = \dot{m}\left[C_j - u\right] \times u$$

$$\text{Power Output} = \frac{1}{2}\dot{m}\left[C_j^2 - u^2\right] \times u$$

$$\eta_p = \frac{\dot{m}\left[C_j - u\right] \times u}{\frac{1}{2}\dot{m}\left[C_j^2 - u^2\right] \times u}$$

$$\eta_p = \frac{2u\left[C_j - u\right]}{C_j^2 - u^2}$$

$$\eta_p = \frac{2u}{C_j + u}$$

Dividing the numerator and denominator by Cj we get,

$$\eta_p = \frac{\frac{2u}{C_j}}{\frac{C_j + u}{C_j}} = \frac{2u / C_j}{1 + u / C_j}$$

$$\eta_p = \frac{2\sigma}{1 + \sigma}$$

Thermal Efficiency

It is defined as the ratio of power output of the engine to the power input of the engine.

$$\eta_f = \frac{\text{Power output of the Engine}}{\text{Power Input the engine through fuel}}$$

$$\text{Power Input} = \dot{m}_f \times \text{C.V.}$$

$$\text{Power Output} = \frac{1}{2}\dot{m}\left[C_P^2 - u^2\right]$$

$$\eta_t = \frac{1/2\dot{m}\left[C_p^2 - u^2\right]}{m_f \times C.V.}$$

If efficiency of combustion is considered,

$$\eta_t = \frac{1/2\dot{m}\left[C_p^2 - u^2\right]}{\eta_B \times mf \times C.V.}$$

Overall Efficiency

It is defined as the ratio of propulsive power to the power input to the engine.

$$\eta_o = \frac{\text{Propulsive Power (or) Thrust Power}}{\text{Power Input the Engine}}$$

We know that,

$$\text{Thrust Power} = m \cdot [Cj - u] \times u$$

$$\text{Power Input} = m_f \times C.V$$

$$\eta_o = \frac{\dot{m}\left[C_j - u\right] \times u}{m_f \times C.V.}$$

$$\eta_o = \eta_p \times C.V.$$

Problems

1. A turbojet engine operating at a Mach number of 0.8 and the altitude is 10 km has the following data. Calorific value of the fuel is 42,800 kJ/kg. Thrust force is 50 kN, mass flow rate of air is 45 kg/s, mass flow rate of fuel is 2.65 kg/s. Let us determine the specific thrust, thrust specific fuel consumption, jet velocity, thermal efficiency, propulsive efficiency and overall efficiency, assuming the exit pressure is equal to ambient pressure.

Solution:

Given:

Mach number M = 0.8

Altitude, Z = 10 km = 10,000 m

Calorific value of fuel, C.V = 42,800 kJ/kg

= 42,800 × 103 J/kg

Thrust force, F = 50 kN = 50×10^3N

Mass of air, ma = 45 kg/s

Mass flow rate of fuel, m_j = 2.65 kg/s

Exit pressure, p_e = Ambient pressure pa

To find:

1. Specific Thrust, FSP.
2. Thrust specific fuel consumption (TSFC).
3. Jet velocity, Cj.
4. Thermal Efficiency, ηt.
5. Propulsive efficiency, ηp.
6. Overall efficiency, ηo.

At Z = 10000 m, from gas tables the properties of air are: T_1 = 223.15 K

ρ_1 = 0.264 bar

ρ1 = 0.413

$$\text{Specific Thrust}, F_{SP} = \frac{F}{\overset{o}{m}}$$

$$\dot{m} = \dot{m}_a + \dot{m}_f$$

$$= 45 + 2.65$$

$$m \quad 47.65\ \text{kg/s}$$

$$F_{Sp} = \frac{50 \times 10^3}{47.65}$$

$$FSP = 1049.31\ \text{N}$$

Thrust Specific fuel consumption,

$$TSFC = \frac{\dot{m}_f}{F} = \frac{2.65}{50000}$$

$$TSFC = 5.3 \times 10^{-5}\ \text{kg/S}-\text{N}$$

$$TSFC = 0.1908\ \text{kg/h r}-\text{N}$$

Jet velocity C_j

We know that,

$$F = \dot{m}C_j - \dot{m}_{au}$$

$$= 47.65 \times C_j - 45 \times (\mu)$$

$$M_1 = \frac{u}{a_1}$$

$$a_1 = \sqrt{\gamma R T_1}$$

$$= \sqrt{1.4 \times 287 \times 223.15}$$

$$a_1 = 299.43 \text{ m/s}$$

$$u = M_1 \times a_1 = 0.8 \times 299.43$$

$$u = 239.548 \text{ m/s}$$

$$F = \dot{m}C_j - \dot{m}_a u$$

$$50 \times 10^3 = 47.65 \times C_j - 45 \times 239.548$$

Velocity of jet $C_j = 1275.54 \text{ m/s}$

Thermal Efficiency,

$$\eta_t = \frac{\frac{1}{2}\overset{o}{m}\left[C_j^2 - u^2\right]}{\overset{o}{m}_f \times C.V}$$

$$= \frac{\frac{1}{2}47.65\left[1275.54^2 - 239.548^2\right]}{2.65 \times 42{,}800 \times 10^3}$$

Thermal Efficiency,

$$\eta_t = 0.3297$$

$$\eta_t = 32.97\%$$

Propulsive efficiency,

$$\eta_p = \frac{2u}{C_j + u}$$

$$= \frac{2 \times 239.548}{1275.54 + 239.543}$$

$$\eta_P = 31.62\%$$

Overall efficiency,

$$\eta_o = \eta_P \times \eta_t$$

$$\eta_o = 0.3161 \times 0.3297$$

$$\eta_o = 0.1042$$

$$\eta_o = 10.42\%$$

Result,

$$FSP = 1049.31$$

$$NTSFC = 0.1908\ kg/hr - N$$

$$C_j = 1275.54\ m/s$$

$$\eta_t = 32.97\%$$

$$\eta_P = 31.62\%$$

2. A turbojet engine operating with its exhaust nozzle unlocked propels an aircraft at a uniform speed of 900 km/h when it develops a thrust of 14kN. The air intake to the engine is 50 kg/s and its air fuel ratio is 85. The calorific value of the fuel supplied to the engine is 44000 kJ/kg. The isentropic enthalpy change in the nozzle is 150kJ/kg. Let us determine the thrust power, propulsive power, propulsive efficiency, thermal efficiency and overall efficiency of the engine.

Solution:

Given:

Aircraft speed,

$$u = 900\ km/h$$

$$= \frac{900 \times 10^3 \text{ m}}{3600}$$

$$= 250 \text{ m/s}$$

Thrust, $F = 14 \text{ kN} = 14 \times 10^3 \text{N}$

Air intake ma 50kg/s

Air fuel ratio, $\frac{m_a}{m_f} = 85$

Calorific Value, $C.V = 44000 \text{kJ/kg}$

$$= 44{,}000 \times 10^3 \text{J/kg}$$

Isentropic enthalpy change, $\Delta h = 150 \text{ kJ/kg}$

$$= 150 \times 10^3 \text{ J/kg}$$

To find:

Thrust power, P

Propulsive power, P

Propulsive efficiency, ηp

Thermal efficiency, ηt

Overall efficiency, ηo

Mass flow rate of air fuel mixture,

$$\dot{m} = \dot{m}_a + \dot{m}_f$$

$$= \dot{m}_a \left[1 + \frac{\dot{m}_f}{\dot{m}_a}\right]$$

$$= 50\left[1 + \frac{1}{85}\right]$$

$$\dot{m} = 50.58 \text{kg/S}$$

Mass flow rate of air fuel mixture,

$$\dot{m} = 50.58 \text{kg/S}$$

$$\dot{m}_f = \overset{o}{m} - \overset{o}{m}_a$$

$$= 50.58 - 50$$

$$\dot{m}_f = 0.58 \text{ kg/s}$$

Thrust power,

$$P = F \times u$$

$$P = 14 \times 10^3 \times 250$$

$$P = 3.50 \times 10^6 \text{ W}$$

Propulsive power,

$$P = F \times u$$

$$P = 14 \times 10^3 \times 250$$

$$= 3.50 \times 10^6 \text{ W}$$

Propulsive efficiency, $\eta_p = \dfrac{2u}{C_j + u}$

$$F = \dot{m}C_j - \dot{m}_a u$$

$$14 \times 10^3 = (50.58 \times C_j) - (50 \times 250)$$

$$C_j = 523.92 \text{ m/s}$$

$$\eta_P = \frac{24}{C_j + u}$$

$$= \frac{2 \times 250}{523.92 + 250}$$

$$\eta_P = 0.64606$$

$$\eta_P = 64.6\%$$

Thermal Efficiency,

$$\eta_t = \frac{\frac{1}{2}\dot{m}\left[C_j^2 - u^2\right]}{\dot{m}_f \times C.U}$$

$$\eta_t = \frac{\frac{1}{2}\times 50.58\left[523.92^2 - 250^2\right]}{0.58\times 44\times 10^6}$$

$$\eta_2 = 0.21008$$

$$\eta_t = 21\%$$

Overall Efficiency,

$$\eta_o = \eta_p \times \eta_t$$

$$= 0.646 \times 0.21$$

$$= 0.13571$$

$$\eta_o = 13.57\%$$

Result,

$$P = 3.50 \text{ x } 10^6 \text{ W}$$

$$\eta_p = 64.6\%$$

$$\eta_t = 21\%$$

$$\eta_o = 13.57\%$$

3.7.1 Turbo Prop

Turbo-Pump Read System used in a Liquid Propellant Rocket

1. In this system liquid fuel and the liquid oxidizer are stored in a separate tank at low pressure.
2. Gas turbine is used to operate the fuel and oxidizer pumps.
3. Liquid hydrogen peroxide [H2CO] from the tank is decomposed by a catalyst such as calcium or sodium permanganate. Due to this steam and oxygen are generated. This steam is used to drive the turbine.

4. Because of the third liquid in the gas turbine pumps, additional lines are necessary. So the pump presumption system is considerably more complex then gas pressurization system.

5. Design of pump is a greatest problem that will handle the liquids safely and without leaks.

6. It is very similar to turbo jet engine. In this type, the turbine drives the compressor and propeller.

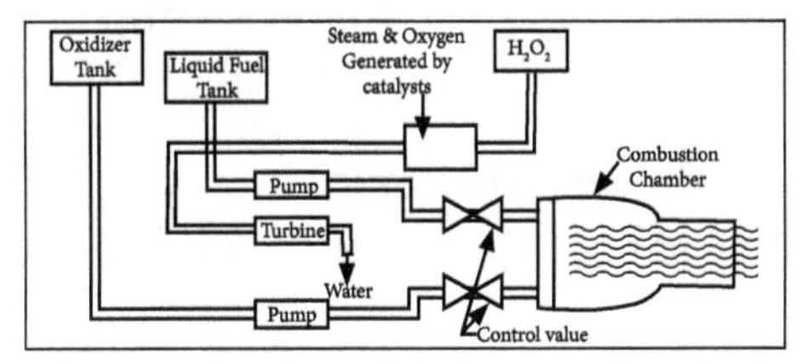

Turbo-Prop Engine or Turbo Propeller Engine.

Construction

1. The construction of Turbo-Prop Engine is shown in the below figure.

2. The function of diffuser is to convert the kinetic energy of the entering air into pressure energy.

3. The function of nozzle is to convert the pressure energy of the combustion gases into kinetic energy.

4. The angular velocity of the shaft is very high. But the propeller cannot run at higher angular velocity. So a reduction gear box is provided before the power is transmitted to the propeller.

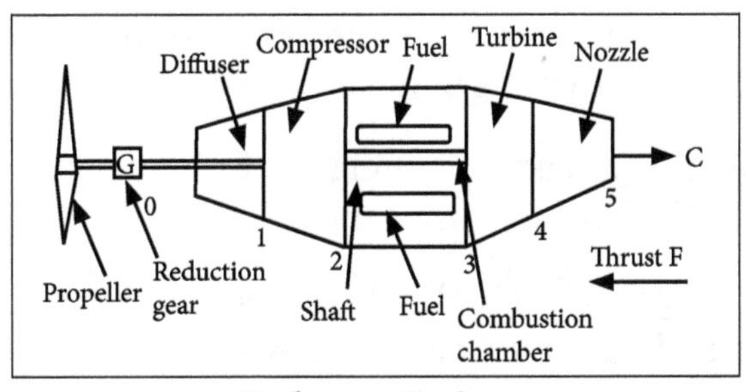

Turboprop Engine.

Turbo-Prop Engine consists of the following:

1. Diffuser.

2. Compressor.

3. Combustion chamber.

4. Turbine.
5. Exhaust nozzle.
6. Reduction year.
7. Propeller.

Working

1. Air from the atmosphere enters into turbo prop engine. The air velocity gets reduced and its static pressure is increased by diffuser.
2. Then the air passes through the rotary compressor in which the air is further compressed. So, the static pressure of the air is further increased.
3. Then the high pressure air flows into the combustion chamber. In the combustion chamber, the fuel is injected by suitable injectors and the air-fuel mixture is burnt. Heat is supplied at constant pressure.
4. The highly heated products of combustion gases then enters the turbine and partially (about 80 to 90%) expanded.
5. The power produced by the turbine is used to drive the compressor and propeller.
6. Propeller is used to increase the flow rate of air which results in better fuel economy.
7. The hot gases from the turbine are then allowed to expand the exhaust nozzle section.
8. In the nozzle, pressure energy of the gas is converted into kinetic energy. So the gases come out from the unit with very high velocity.
9. Due to high velocity of gases coming out from the unit, a reaction (or) thrust is produced in the opposite direction.
10. The total thrust produced in this engine is the sum of the thrust produced by the propeller and the thrust produced by the nozzle. This total thrust propels the air craft.

Advantages

1. High take-off thrust.
2. Good propeller efficiency at a speed below 800 km/hr.
3. Reduce vibration and noise.

4. Better fuel economy.
5. Easy maintenance.
6. It operates over a wide range of speeds due to multi shaft arrangement.
7. The power output is not limited.
8. Sudden decrease of speed is possible by thrust reversal.

Disadvantages

1. The propeller efficiency rapidly decreases at high speeds due to shocks and flow separation.
2. It requires a reduction gear which increases the cost of the engine.
3. More space needed than turbo jet engine.
4. Engine construction is more complicated.

Applications

The turbo prop engine is best suited for commercial and military air-craft operation due to its high flexibility of operation and good fuel economy.

3.7.2 Turbo Fan

Turbo Fan

The turbofan engine is a combination of the turbo prop and the turbojet engines combining the advantages of both.

Construction

The construction of turbofan is shown in the below figure:

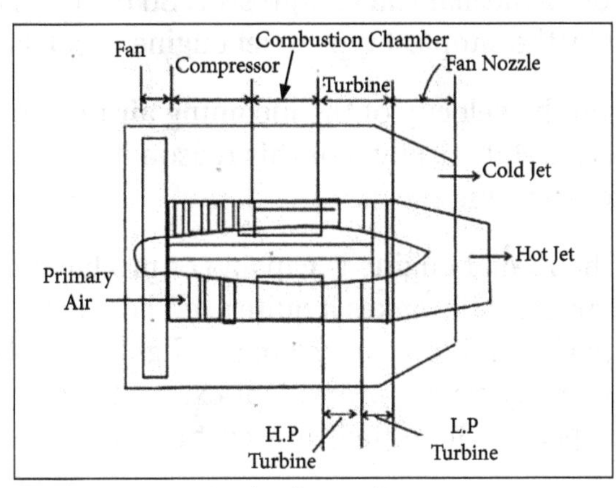

Turbo fan engine.

Working

1. Air from the atmosphere enters in to turbo fan engine employing a low pressure ducted fan.

2. The air after passing through the fan is divided in to the streams, namely primary air and secondary air.

3. The primary air (m_h) flow through the turbofan engine consisting of compressor, combustion chamber, turbine and exhaust nozzle. Combustion takes place in the combustion chamber and the thrust is produced in the opposite direction.

4. The secondary air (or) by pass air (or) cold air (m_e) at relatively lower pressure flows around the turbofan engine and expands in the fan nozzle. Hence thrust is produced.

5. The thrust developed by the secondary air is at lower velocity and the thrust developed by the primary air is at much higher velocity.

6. The total thrust produced in this engine is the sum of thrust produced by the primary air (m_h) and the secondary air (m_c). This total thrust propels the air craft.

7. The ratio of the mass flow rates of cold air (m_c) and the hot air (m_h) is known as By pass ratio.

3.7.3 Ram Jet Engines

The ramjet engine is an air breathing engine which operates on the same principle as the turbojet engine. Its basic operating cycle is similar to that of the turbojet. It compresses the incoming air by ram pressure, adds the heat energy to velocity and produces thrust. The converting kinetic energy of the incoming air into pressure, the ramjet is able to operate without a mechanical compressor. So the engine requires no moving parts and is mechanically the simplest type of jet engine which has been devised.

Therefore, it depends on the velocity of the incoming air for the needed compression, the ramjet will not operate statistically. For this reason it requires a turbojet or rocket assist to accelerate it to operating speed.

At supersonic speeds the ramjet engine is capable of producing very high thrust with high efficiency. This characteristic makes it quite useful on high speed aircraft and missiles, where its great power and low weight make flight possible in regions where it would be impossible with any other power plant except the rocket. Ramjets have also been used at subsonic speeds where their low cost and light weight could be used to advantage.

Principle of Operation

The ramjet consists of a diffuser, flame holder, fuel injector, combustion chamber and exit nozzle As shown in figure (a). The air taken in by the diffuser is compressed in two stages. The external compression takes place because the bulk of the approaching engine forces the air to change its course. Further compression is accomplished in the diverging section of the ramjet diffuser. Fuel is injected into and mixed with air in the diffuser.

The flame holder provides a minimum velocity region favourable to flame propagation and the fuel-air mixture recirculates within this sheltered area and ignites the fresh charge as it passes the edge of the flame holder.

The burning gases then pass through the combustion chamber, increasing in temperature and therefore in volume. Because of the volume of air increases, it must speed up to get out of the way off the fresh charge following behind it and a further increase in velocity occurs as the air is squeezed out through the exit nozzle. The thrust produced by the engine is proportional to this increase in velocity.

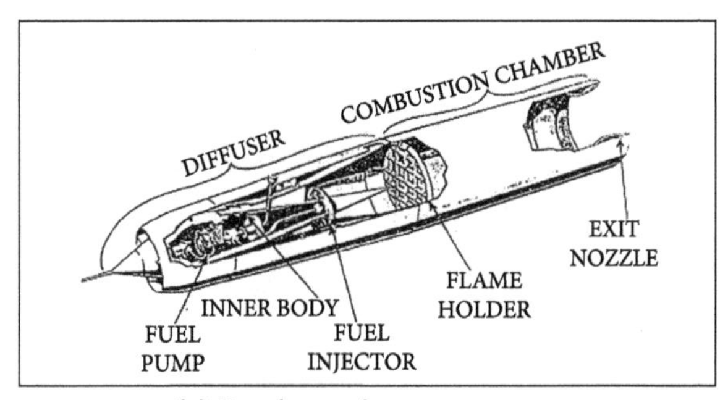

(a) Ramjet engine component.

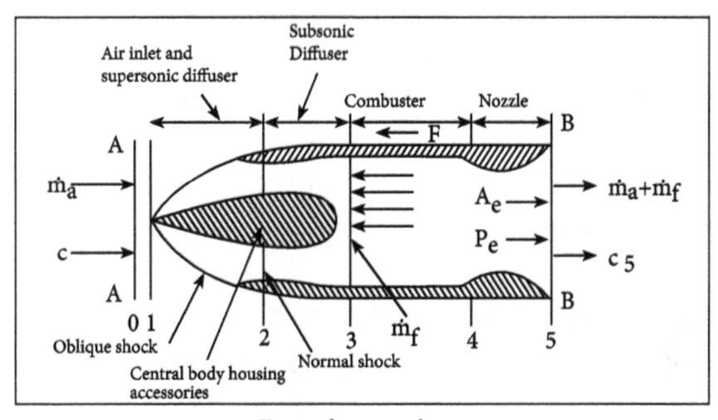

Ram jet engine.

Construction

The construction of ramjet engine is shown in the above figure which is simplest types of air-breathing engine. It consists of:

1. Supersonic diffuser (1-2).
2. Subsonic diffuser (2-3).

3. Combustion chamber (3-4).

4. Discharge nozzle section (4-5).

The function of supersonic and subsonic diffusers is to convert the kinetic energy of the entering air into pressure energy. This energy transformation is called ram effect and the pressure rise is called the ram pressure.

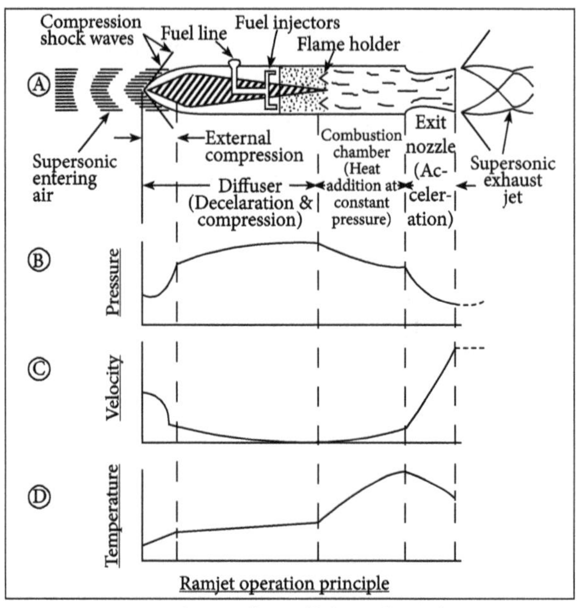

Operation and condition of ramjet.

Working

1. Air from the atmosphere enters the engine at a very high speed and its velocity gets reduced and its static pressure is increased by supersonic diffuser.

2. Then the air passes through the subsonic diffuser and its velocity further reduces to subsonic value.

3. Then the high pressure air flows into the combustion chamber. The highly heated products of combustion chamber gases are then allowed to expand.

4. The air entering the engine with a supersonic speed must be reduced to subsonic speed. This is necessary to prevent the blow out of the flame in the combustion chamber. The velocity must be small enough to make it possible to add the required quantity of fuel for stable combustion.

5. Both theory and experiment indicate the speed of the air entering the combustion chamber should not be higher than that corresponding to a local match number of 0.2 approximately.

Advantages

1. Since turbine is not used the maximum temperature which can be allowed in ramjet is very high, about 2000°C as compared to about 1000°C in turbojets. This allows a greater thrust to be obtained by burning fuel at A/F ratio of about 15.1 which gives higher temperatures.
2. Ramjet is very simple and does not have any moving part. It is very cheap and requires almost no maintenance.
3. There seem to be no upper limit to the flight speed of the ramjet.
4. The SFC is better than turbojet engines at high speed and high altitudes.

Disadvantages

1. The engine heavily relies on the diffuser and it is very difficult to design a diffuser which will give good pressure recovery over a wide range of speeds.
2. Since the compression of air is obtained by virtue of its speed relative to the engine, the take-off thrust is zero and it is not possible to start a ramjet without an external launching device.
3. At very high temperature of about 2000° C dissociation of products of combustion occurs which will reduce the efficiency of the plant if not recovered in nozzle during expansion.
4. Due to high air speed, the combustion chamber requires flame holder to stabilize the combustion.

Application

1. Subsonic ramjets are used in target weapons, in conjunction with turbojets or rockets for getting the starting torque.
2. Due to its high thrust at high operational speed, it is widely used in high speed air crafts and missiles.

Problems

1. A ramjet engine propels an aircraft at a Mach number of 1.4 and at an altitude of 6000 m. The diameter of the inlet diffuser at entry is 40 cm and the calorific value of the fuel is 43 MJ/kg. The stagnation temperature at the nozzle entry is 1500 K. The properties of the combustion gases are same as those of air. ($\gamma = 1.4$, $R = 287$ J/kg K). Let us determine the following:

i. The efficiency of the ideal cycle.

ii. Flight speed.

iii. Air flow rate.

iv. Diffuser pressure ratio.

v. Fuel air ratio.

vi. Nozzle pressure ratio.

vii. Nozzle jet Mach number.

viii. Propulsive efficiency.

ix. Thrust.

Solution:

Assuming diffuser efficiency, $\eta_D = 0.92$

Combustion efficiency, $\eta B = 0.97$ and Nozzle jet efficiency, η_n (or) $\eta_j = 0.95$. Stagnation pressure loss in the combustion chamber = $0.02\ Po_2$.

Given:

$$M_1 = 1.4$$

$$\eta_d = 0.92$$

$$Z = 6000 \text{ m}$$

$$\eta_B = 0.93$$

$$d_1 = 40 \text{ cm}$$

$$\eta_N \left(\text{or}\right) \eta_j = 0.95$$

$$C_v = 43 \text{ MJ/kg}$$

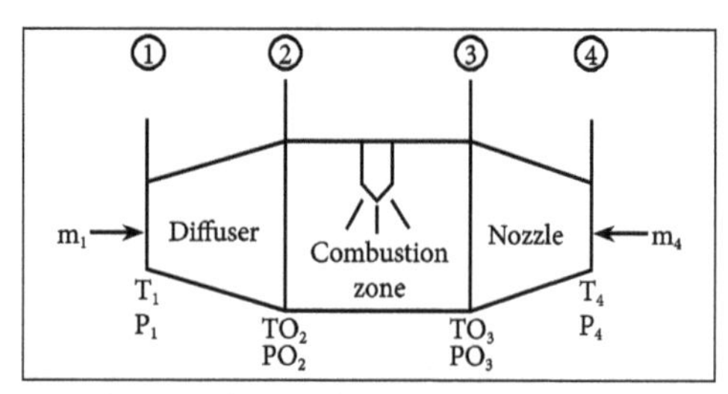

Pressure loss in the combustion chamber = $0.02\ P_{02}$.

$$T_o = 1500 \text{ K}$$

$$\eta_1 = \frac{1}{1+\frac{1}{\gamma-1}\times 1/m_1^2}$$

$$= \frac{1}{1 + \frac{2}{1.4 - 1} \times \frac{1}{(1.4)^2}}$$

$$\eta_1 = 0.281$$

$$\eta_1 = 28.1\%$$

Refer gas tables at Z = 6000 m (Gas table Page No. 1).

$$p_i = 0.660 \ kg/m^3 \ T_i = 249.15 \ k$$

$$a_i = 316.5 \ m/sp_i = 0.472 \ bar$$

$$M_1 = u/a_1$$

$$u = M_1 \times a_1$$

$$u = 443.10 \ m/s$$

Area of cross section,

$$A_1 = \frac{\pi}{4} d_1^2$$

$$= \frac{\pi}{4}(0.40)^2$$

$$A_1 = 0.125 \ m^2$$

$$\dot{m}_a = P_1 A_1 u$$

$$= 0.660 \times 0.125 \times 443.10$$

$$m_a = 36.55 \ kg/s$$

Diffuser efficiency,

$$\eta_b = \frac{(R_{ob})^{\frac{\gamma-1}{\gamma}} - 1}{\frac{\gamma-1}{2} M_1^2}$$

$$0.92 = \frac{\left(R_{ob}\right)^{\frac{1.4-1}{1.4}} - 1}{\frac{1.4-1}{2} \times (1.4)^2}$$

$$R_{ob} = \frac{P_{02}}{P_1} = 2.938$$

Stagnation Temperature – Mach number Relation,

$$\frac{T_o}{T} = 1 + \frac{\gamma - 1}{2} M^2$$

$$\frac{T_{01}}{T} = 1 + \frac{\gamma - 1}{2} M_1^2$$

$$T_{01} = 346.81\,K = T_{02}$$

$$\eta_B = \frac{m_a^o . C_P \left(T_{03} - T_{02}\right)}{m_f \times C.V.}$$

$$0.97 = \frac{36.55 \times 1005 (1500 - 346.81)}{m_f \times 43 \times 10^6}$$

$$\dot{m}_f = 1.015 \text{ kg/s}$$

$$\frac{\dot{m}_f}{\dot{m}_a} = \frac{1.015}{36.55} = 0.027$$

$$\text{Fuel air ratio } f = \frac{\dot{m}_f}{\dot{m}_a} = 0.027$$

$$P_{03} = P_{02} - 0.02 P_{02}$$

$$= P_{02}(1 - 0.02)$$

$$P_{03} = 0.98 P_{02}$$

Diffuser Pressure ratio,

$$R_{oD} = \frac{}{}{}^{02} = 2.938$$

$$P_{02} = 2.938 \times P_1$$

$$P_{02} = 2.938 \times 0.472 \times 108^1$$

$$P_{02} = 1.386 \times 10^5 \, N$$

$$P_{03} = 0.98 \times 1.386 \times 10^5$$

$$P_{03} = 1.359 \times 10^5 \, N/m^2$$

Nozzle Pressure Ratio,

$$R_{ON} = R_o, = \frac{P_{03}}{P_4} = \frac{1.359 \times 10^5}{0.472 \times 10^5}$$

$$R_{ON} = 2.87,$$

$$\frac{P_4}{P_{03}} = \frac{1}{2.87} = 0.348$$

Refer Isentropic flow table for y = 1.4 & $\frac{P_9}{P_{03}} = 0.348$

$$M_{4-5} = 1.325 [\text{Page No. } 32]$$

$$\frac{T_{45}}{T_{04}} = 0.740, \dot{T}_{45} = 1110 K$$

Nozzle efficiency,

$$\eta_N = \frac{T_{04} - T_4}{T_{04} - T_{45}}$$

$$0.95 = \frac{1500 - T_4}{1500 - 1110}$$

$$T_4 = 1129.5 \, K$$

$$T_o = T + (C^2/2) \, C_P$$

$$T_{04} = T + (C_4^{\,2}/2) \, C_P$$

$$1500\times1129.5+\frac{C_4^2}{2}\times1005$$

$$C_4^2 = 744.90\times10^3 \text{ m/s}$$

$$C_4 = 862.96 \text{ m/s}$$

$$a_4 = \sqrt{\gamma RT_4}$$

$$= \sqrt{1.4\times187\times1129.5}$$

$$a_4 = 673.67 \text{ m/s}$$

Nozzle Jet Mach number (or) Mach number at exit (M4),

$$M_4 = C_4 / a_4$$

$$= \frac{862.96}{673.67}$$

$$M_4 = 1.280$$

Propulsive efficiency is given by,

$$\eta_P = \frac{24}{C_4+4}$$

$$\eta_P = \frac{2\times443.10}{862.96+44.10}$$

$$\eta_P = 67.8\%$$

Mass flow rate of air-fuel mixture,

$$m = m_a + m_f$$

$$= 36.55 + 1.1015$$

$$m = 37.65 \text{ kg/s}$$

$$\text{Thrust } F = m_4 c - m_4 a$$

$$= (37.65) \times 862.96 - 36.55 \times 443.10$$

$$F = 66.22\times10^3 \text{ N}$$

3.8 Axial Flow and Centrifugal Compressor

Axial Flow Compressors: Mechanical Details and Principle of Operation

In axial flow compressor the air flows parallel to the axis. It consists of a number of rotating blades fixed to a rotating drum. Each stage consists of one row of moving blades and one row of fixed blades. As the air flows from one set of stator and rotor to another, it gets compressed. The successive compression of air in all the sets of stator and rotor increases the pressure and the air is delivered at a high pressure from the outlet.

The blades are made of aero foil section to reduce the turbulence loss and boundary separation. The number of stages varies from 4 to 16. The pressure ratio per stage varies between 1.2 to 1.3.

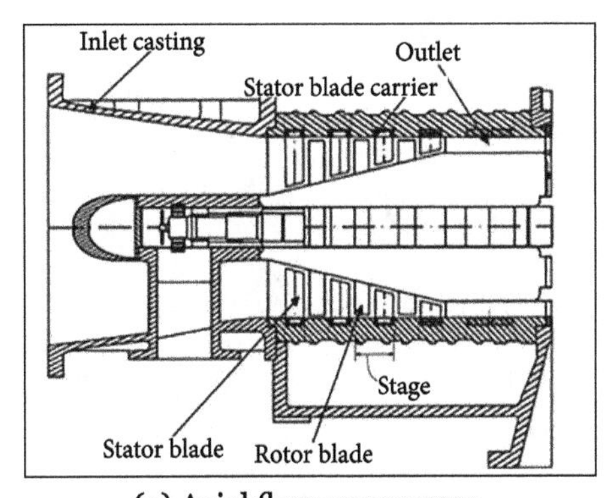

(a) Axial flow compressor.

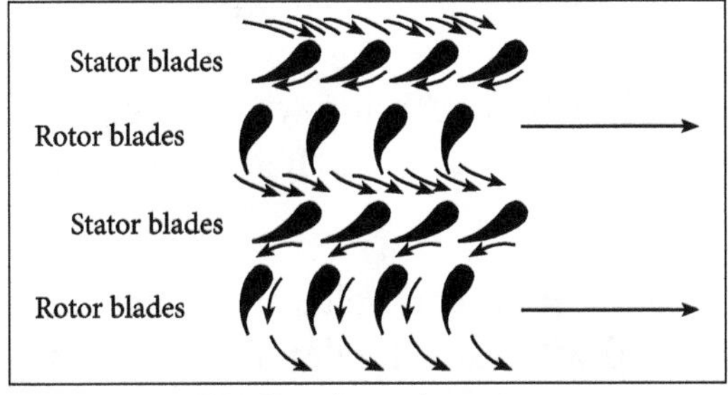

(b) Flow through stages.

The basic principle of acceleration of the working fluid, followed by diffusion to convert acquired kinetic energy into a pressure rise, is applied in the axial compressor. The flow is considered as occurring in a tangential plane at the mean blade height where the blade peripheral velocity is U.

This two dimensional approach means that in general the flow velocity will have two components, one axial and one peripheral denoted by subscript w, implying a whirl velocity. It is first assumed that the air approaches the rotor blades with an absolute velocity, V_1 at an angle α_1 to the axial direction. In combination with the peripheral velocity U of the blades, its relative velocity will be V_{r1} at an angle β_2 as shown in the upper velocity triangle in above the figure.

Advantages

1. More efficiency.
2. Straight-through flow, allowing high ram efficiency.
3. Small frontal area for given airflow.
4. Increased pressure rise due to increased number of stages with negligible losses.

Disadvantages

1. A good efficiency over narrow rotational speed range.
2. Relatively high weight.
3. Difficulty of manufacture and high cost.
4. High starting power requirements (this has been partially overcome by split compressors).

Centrifugal Compressor

The centrifugal compressors are also called as radial compressor. In general the centrifugal compressor is suitable to handle large quantity of air with moderate pressure ratio of the order of 4 (or) 6:1. The basic components of a typical centrifugal compressor include an impeller rotating within a casing. The impeller consists of a disc onto which radial blades are attached. The impeller is shrouded by the casing.

The blades split the air, taken into the impeller into cells. As the impeller rotates, the cells of air will also be rotated with the impeller. Due to centrifugal force the air in the cells will move out towards the outside edge of the impeller and more fresh air will move into the centre of the impeller as a new charge.

The centre of the impeller is called as eye. As the air passes away from the outside edge of the impeller, the air passes into a diffuser ring which helps to direct it into the volute. Due to increased area of the volute casing from the outside edge of impeller air gets decelerated producing a rise in pressure.

The volute can be considered as a collecting device for the compressor. A duct leads away from the volute to take the compressed air out of the compressor. The eye of the impeller is connected with drive shaft as shown in the figure.

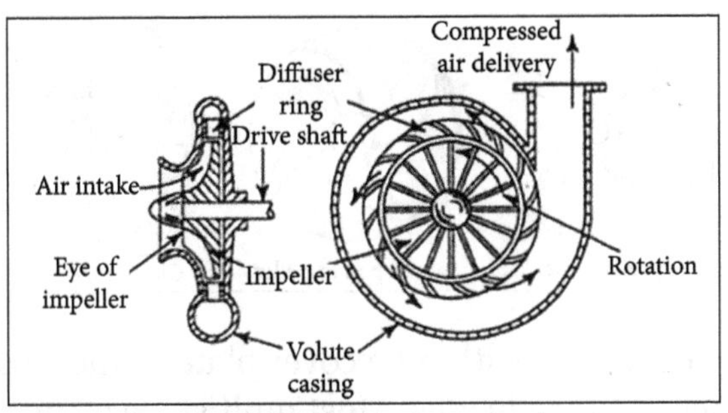

Centrifugal Compressor.

Variations of Pressure and Velocity

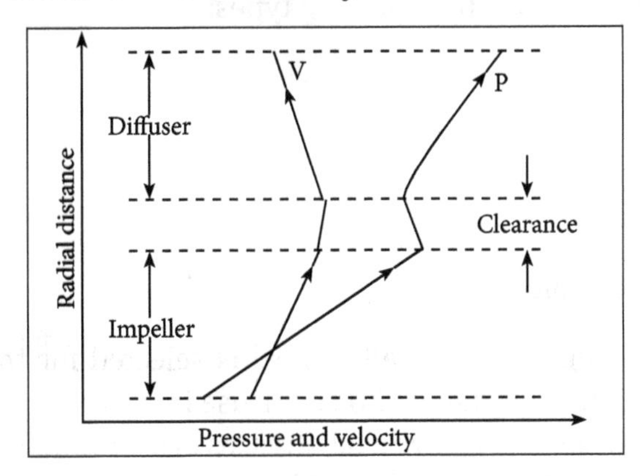

The figure shows the variation of the pressure and the velocity across the impeller and the diffuser. As the fluid approaches the impeller, it is subjected to centrifugal effect thereby the kinetic energy and the pressure of the fluid both increases along the radial direction.

When the impeller discharges the fluid into the diffuser, the static pressure of the fluid rises due to the deceleration of the flow. Hence the velocity reduces and the pressure still increases as shown in Figure. This is mainly due to the conversion of kinetic energy into pressure energy of the fluid.

Impeller Blade Shape

The impellers impart velocity to the gas with blades that are attached to a rotating disc. The impeller blades are forward-leaning, radial or backward-leaning depending on the desired performance characteristic curve. Backward-leaning blades tend to provide the widest operating range with good efficiency.

They are the generally used blade shape. Proper sizing of the impeller flow channels is determined by the volumetric flow rate to control gas velocities through the impeller. In a multistage compressor, the impellers should be properly sized for peak performance and properly matched to accommodate the volumetric flow rate reduction through the compressor.

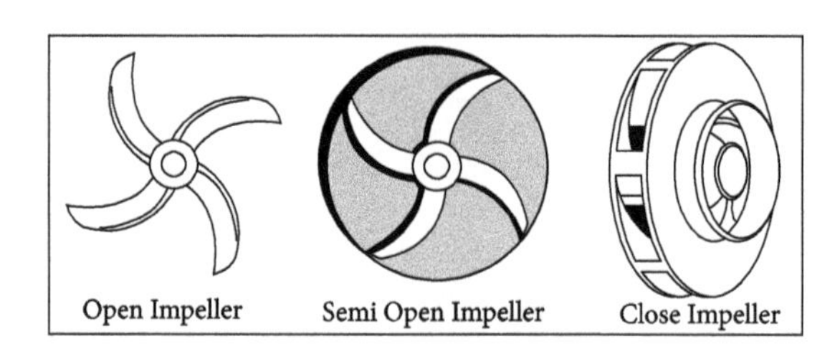

Impellers can be of the open type without a cover plate or the closed type that incorporates a cover plate attached to the blades. Most multistage compressors use the closed-type impeller design.

Impeller construction can be of the following types:

1. Riveted.
2. Brazed.
3. Electron beam welded.
4. Welded conventionally.

For most applications, high-strength alloy steel is selected for the impeller material. Stainless steel is often the material of choice for use in corrosive environments. As the impellers rotate at high speeds, centrifugal stresses are an important design consideration and high-strength steels are used for this purpose.

The losses in a centrifugal compressor are as follows:

Frictional Losses

A major portion of the losses occurs due to fluid friction in stationary and rotating blade passages. The flow in impeller and diffuser are decelerating in nature. Therefore the frictional losses are due to both skin friction and boundary layer separation. These losses depend on the friction factor, length of the flow passage and square of the fluid velocity.

Incidence Losses

During the off-design conditions, the direction of relative velocity of fluid at inlet does not match with the inlet blade angle and therefore fluid cannot enter the blade passage smoothly by gliding along the blade surface. The loss in energy that takes place because of this is known as incidence loss (also known as shock losses).

Clearance and Leakage Losses

The minimum clearances are necessary between the impeller shaft and the casing between the outlet periphery of the impeller eye and the casing. The leakage of gas through the shaft clearance is minimized by employing glands.

These losses depend upon the impeller diameter and the static pressure at the impeller tip. A larger diameter of impeller is necessary for a higher peripheral speed U2 and it is very difficult in the situation to provide sealing between the casing and the impeller eye tip.

The leakage losses comprise a small fraction of the total loss. The incidence losses attain the minimum value at the designed mass flow rate.

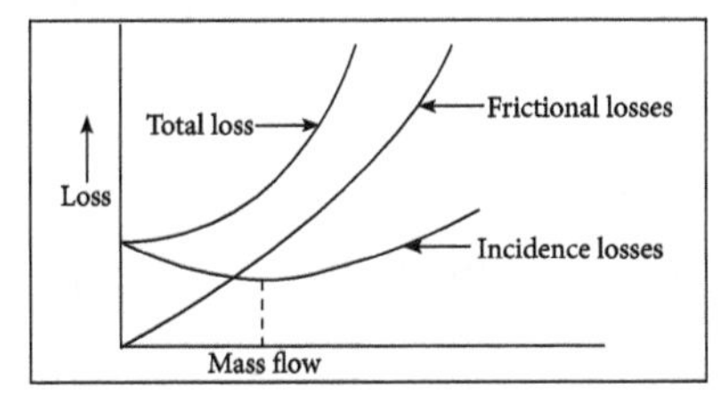

Various losses in a centrifugal compressor.

The shock losses are zero at the designed flow rate. The incidence losses comprises both shock losses and impeller entry loss due to a change in the direction of fluid flow from axial to radial direction in the vaneless space before entering the impeller blades.

The impeller entry loss is similar to that in a pipe bend and is very small compared to other losses. Hence the incidence losses show a non-zero minimum value at the designed flow rate.

Power Input Factor

The power input factor takes into account the effect of disk friction, windage, etc. Considering all these losses, the actual work done (or energy input) on the air per unit mass becomes,

$$w = \Psi \sigma U_2^2 \qquad ...(1)$$

where Ψ is the power input factor. From steady flow energy equation and air as an ideal gas, one can write for adiabatic work w per unit mass of air flow as,

$$w = c_p\left(T_{0_2} - T_{0_1}\right) \qquad ...(2)$$

where T_{01} and T_{02} are the stagnation temperatures at inlet and outlet of the impeller and Cp is the mean specific heat over the entire temperature range.

$$w = \Psi \sigma U_2^2 = c_p\left(T_{0_2} - T_{0_1}\right) \qquad ...(3)$$

The stagnation temperature represents the total energy held by a fluid. Since no energy is added in the diffuser, the stagnation temperature rise across the impeller must be equal to that across the whole compressor. If the stagnation temperature at the outlet of the diffuser is designated by T_{03}, then $T_{03} = T_{02}$. One can write from equation (3).

$$\frac{T_{0_2}}{T_{0_1}} = \frac{T_{0_3}}{T_{0_1}} = 1 + \frac{\Psi \sigma U_2^2}{c_p T_{0_1}} \qquad ...(4)$$

The overall stagnation pressure ratio can be written as,

$$\frac{p_{0_3}}{p_{0_1}} = \left(\frac{T_{0_{3s}}}{T_{0_1}}\right)^{\frac{\gamma}{\gamma-1}}$$

$$= \left[1 + \frac{\eta_c\left(T_{0_3} - T_{0_1}\right)}{T_{0_1}}\right]^{\frac{\gamma}{\gamma-1}} \qquad ...(5)$$

Where, T_{03s} and T_{03} are the stagnation temperatures at the end of an ideal (isentropic) and actual process of compression respectively and ηc is the isentropic efficiency defined as,

$$\eta_c = \frac{T_{0_{3s}} - T_{0_1}}{T_{0_3} - T_{0_1}} \qquad ...(6)$$

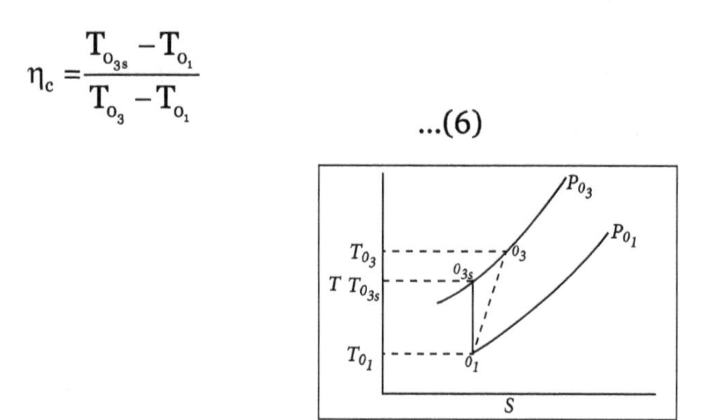

Ideal and actual processes of compression on T-s plane.

Since the stagnation temperature at the outlet of impeller is same as that at the outlet of the diffuser, one can also write T_{02} in place of T_{03} in equation (6). The power input factor lie in the region of 1.035 to 1.04. By knowing ηc we can calculate the stagnation pressure rise for a given impeller speed. The relative velocity Vγ1 at the eye tip has to be held low otherwise the Mach number (based on $V_{\gamma 1}$) given by $M_{\gamma 1} = \frac{V_{\gamma 1}}{\sqrt{\gamma R T_1}}$ will be too high causing shock losses. Mach number Mγ1 should be in the range of 0.7-0.9.

The Head Coefficient

The head coefficient is another dimensionless performance parameter that expresses the actual head as the fraction of the maximum theoretical head at zero flow for a given impeller tip speed and geometry with radial inlet. The theoretical head, expressed in ft-lb per lb or kg-meter per kg is obtained by using the relationship,

$$H_1 - (V_{2w}.U_2 - V_{1n}.U_1)/g$$

The maximum theoretical head is thus equal to U_2^2/g. Therefore,

Head coefficient $\Psi = H/\left(U_2^2\right)/g$,

The incremental reduction from a calculated theoretical head is due to eddy current and separation losses at the entrance and exit of the impeller. These are explained as obstructions in the path of gas travel and consequent changes of velocity and direction.

In the case of radial discharge β_2 - 900, theoretically. the head does not vary with load and the head versus flow (H - Q) characteristic should simply be a line parallel to Q on the x-axis of the H versus Q plot. The vane angle at the discharge is inclined so as to represent a backward lean, as seen in most of the applications is (i.e.. $\beta_2 < 90$").

Degree of Reaction

It is defined as the ratio of pressure rise in the impeller to the pressure rise in the compressor. The pressure rise in the impeller is due to the change in kinetic energy due to relative velocity.

$$= \frac{V_{r1}^2 - V_{r2}^2}{2} + \frac{u_2^2 - u_1^2}{2}$$

Pressure rise in the compressor is equal to work done in the compressor.

Hence, pressure rise in compressor = Vw_2u_2

Degree of reaction (R_d)

$$= \frac{\frac{V_{r1}^2 - V_{r2}^2}{2} + \frac{u_2^2 - u_1^2}{2}}{V_{w2}u_2}$$

$$\therefore R_d = \frac{\left(V_{r1}^2 - V_{r2}^2\right) + \left(u_2^2 - u_1^2\right)}{2V_{w2}u_2}$$

From inlet velocity triangle,

$$V_{r1}^2 - u_1^2 = V_{f1}^2 - V_{f2}^2 \left(\text{since } v_{f1} = v_{f2}\right)$$

From outlet velocity triangles,

$$V_{f2}^2 + \left(u_2 - V_{w2}\right)^2 = V_{r2}^2$$

$$V_{r2}^2 = V_{f2}^2 + u_2^2 + V_{w2}^2 - 2u_2V_{w2}$$

$$\therefore u_2^2 - V_{r2}^2 = 2u_2V_{w2} - V_{f2}^2 - V_{w2}^2$$

Now,

$$R_d = \frac{\left(V_{r1}^{\ 2} - V_{r2}^{\ 2}\right) + \left(u_2^{\ 2} - u_1^{\ 2}\right)}{2V_{w2}u_2}$$

It can be modified as,

From the earlier explanation,

$$V_{r1}^{\ 2} - u_1^{\ 2} = V_{f2}^{\ 2}$$

$$u_2^{\ 2} - V_{r2}^{\ 2} = 2u_2V_{w2} - V_{f2}^{\ 2} - V_{w2}^{\ 2}$$

$$\therefore R_d = \frac{V_{f2}^{\ 2} + 2u_2V_{w2} - V_{f2}^{\ 2} - V_{w2}^{\ 2}}{2V_{w2}u_2}$$

$$R_d = 1 - \frac{V_{w2}}{2u_2}$$

Velocity Triangle

In the centrifugal compressor, air enters radially and leaves the impeller axially. The inlet velocity triangles and exit velocity triangles can be drawn on the common base, since the flow velocity remains constant at inlet and outlet. Since the fluid enters radially, the air angle at impeller inlet is equal to 900 (α = 900).

Procedure

1. Draw a vertical line AB to represent the flow velocity and it remains constant at inlet and exit.
2. The horizontal CA represent the blade speed at inlet.
3. The line CB inclined at an angle of fit, blade angle at inlet, is the relative velocity of blade at inlet.

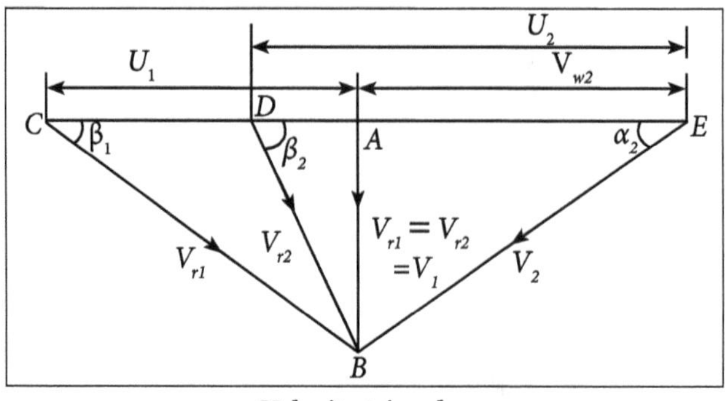

Velocity triangle.

4. Draw line BD inclined at β2, blade angle at outlet to represent the relative blade speed at outlet.

5. From D mark E to represent the blade speed at the outlet of the impeller.

6. Join EB, results in leaving velocity of air V_2 inclined at α_2 with respect to horizontal.

From the Velocity Triangles,

W/D by the compressor per kg of working substance

$$W/D = V_{w1}u_1 + V_{w2}u_2$$

Since, $V_{w1} = 0$,

$$W/D = V_{w2} \cdot u_2$$

Power developed=m×W/D,

$$= m.\ V_{w2} \cdot U_2$$

Velocity Triangles: Energy Transfer Per Stage Degree of Reaction and Work Done Factor

The velocity passing through the diverging passages formed between the rotor blades which do work on the air and increase its absolute velocity, the air will emerge with the relative velocity of V_{r2} at angle β_2 which is less than β_1.

This turning of air towards the axial direction is, as previously mentioned, necessary to provide an increase in the effective flow area and is brought about by the camber of the blades. Since V_{r2} is less than V_{r1} due to diffusion, some pressure rise has been accomplished in the rotor. The velocity V_{r2} in combination with U gives the absolute velocity V_2 at the exit from the rotor at an angle α2 to the axial direction.

The air then passes through the passages formed by the stator blades where it is further diffused to velocity V_3 at an angle α_3 which in most designs equals to α1 so that it is prepared for entry to next stage. Again the turning of the air towards the axial direction is brought about by the camber of the blades.

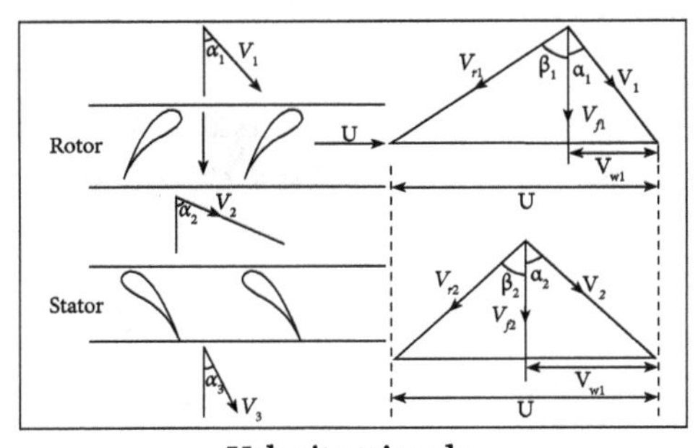

Velocity triangle.

From the above velocity diagram, we get:

$$\frac{U}{V_f} = \tan\alpha_1 + \tan\beta_1 \qquad ...(1)$$

$$\frac{U}{V_f} = \tan\alpha_2 + \tan\beta_2 \qquad ...(2)$$

In which $V_f = V_{f1} = V_{f2}$ is the axial velocity, assumed constant through the stage. The work done per unit mass or specific work input, w being given by,

$$W = U(V_{w2} - V_{w1}) \qquad ...(3)$$

This expression can be put in terms of the axial velocity and air angles to give,

$$w = UV_f(\tan\alpha_2 - \tan\alpha_1) \qquad ...(4)$$

From equation 1 and 2 we get,

$$w = UV_f(\tan\beta_1 - \tan\beta_2) \qquad ...(5)$$

The input energy will be used in raising the pressure and velocity of the air. A part of it will be spent in overcoming the losses due to friction. The input will reveal itself as a rise in the stagnation temperature of the air ΔT_o.

If the absolute velocity of the air leaving the stage V3 is made equal to the entry velocity V1, the stagnation temperature rise ΔT_o will also be the static temperature rise of the stage ΔT_s so that,

$$\Delta T_o = \Delta T_s = UV_f / C_p(\tan\beta_1 - \tan\beta_2) \quad ...(6)$$

The stage temperature rise will be less than that given in equation (6) owing to three dimensional effects in the compressor annulus. It is necessary to multiply the right hand side of equation (6) by the work-done factor λ which is a number less than unity. This is a measure of the ratio of actual work-absorbing capacity of the stage to its ideal value.

The radial distribution of axial velocity is not constant across the annulus but becomes increasingly peaky as the flow proceeds, settling down to a fixed profile at about the fourth stage. Equation (5) can be written with the help of equation (1) as,

$$\begin{aligned} w &= U\left[(U - V_f\tan\alpha_1) - V_f\tan\beta_2\right] \\ &= U\left(U - V_F(\tan\beta_1 - \tan\beta_2)\right) \end{aligned} \qquad ...(7)$$

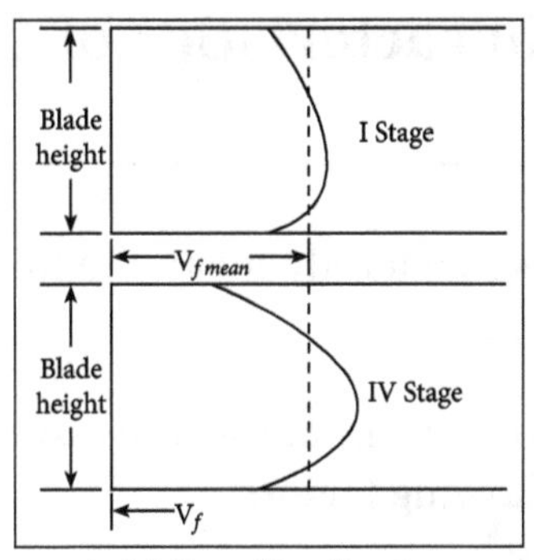

Axial flow distribution.

Since, the outlet angles of the stator and the rotor blades are fixed with the value of α_1 and β_2 and hence the values of ($\tan \alpha_1 + \tan \beta_2$) are also fixed. An increase in V_f will result in a decrease in w and vice-versa.

If the compressor is designed for constant radial distribution of Vf, increase in Vf in the central region of the annulus will reduce the work capacity of blading. This reduction is compensated by an increase in w in the regions of the root and tip of the blading because of the reduction of Vf at these parts of the annulus. The net result is a loss in total work capacity and this effect increases as the number of stages is increased.

If w is the expression for the specific work input (Equation.3), then λw is the actual amount of work which can be supplied to the stage. The application of an isentropic efficiency to the resulting temperature rise will yield the equivalent isentropic temperature rise. The actual stage temperature rise is given by,

$$\Delta T_o = \frac{\lambda U V_f}{C_p}(\tan\beta_1 - \tan\beta_2) \qquad ...(8)$$

And the pressure ratio Rs by,

$$R_s = \left[1 + \frac{n_s \Delta T_o}{T_{01}}\right]^{\frac{\gamma-1}{\gamma}} \qquad ...(9)$$

Where, T_{01} is the inlet stagnation temperature and ηs is the stage isentropic efficiency.

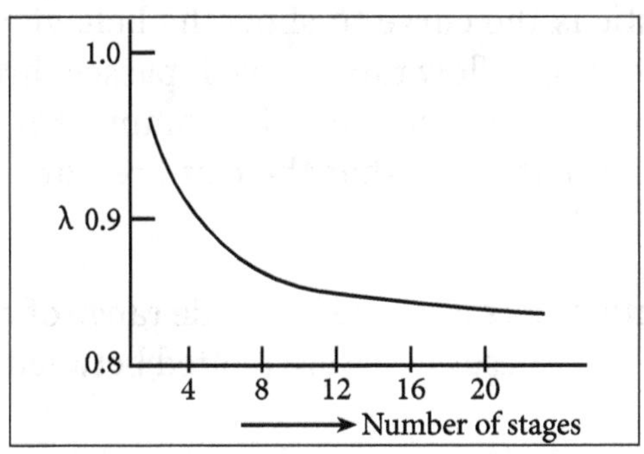

Variation of work-done factor with number of stages.

3.9 Performance Characteristics of Centrifugal and Axial Flow Compressor

By compressor performance, we generally mean the mass of air delivered per minute per B.P. or B.H.P. on the machine.

For a machine of given capacity and numerical pressures the performance of a compressor is influenced by the following factors:

1. The pressure range per cylinder.
2. The number of stages employed.
3. The clearance volume.
4. The speed of the machine.
5. The cooling efficiency.
6. The air intake piping.
7. The type and disposition of the valves.

Effect of Atmospheric Conditions on the Output of a Compressor

A low barometer and a high temperature (as encountered at considerable elevations during day time in tropical countries) is responsible for an appreciable diminution in the mass output of compressors which have to operate under these conditions.

The volumetric efficiency (when referred to a standard atmosphere) falls by about 3% per 300 mm increase in elevation and 1% per 5°C increase in temperature. As a result of the considerable reduction in temperature after sun-down and accompanying humidity, power plant in tropical climates runs considerably better at night.

Compressor characteristic is the curve to show the behavior of fluid like change in pressure, temperature, entropy, flow rate etc. as it passes through the compressor at different compressor speeds. The function of a compressor is to increase the pressure of a fluid passing through it, so that the exit pressure is higher than the inlet pressure.

Due to this property, compressors are used in a wide range of machines, such as refrigerators, cars and jet engines. These curves are plotted between various parameters and some are as follows.

3.9.1 Effects of Slip and Surging

Slip and Slip Factor

A centrifugal compressor has maximum work when $V_{w2} = u_2$. But in actual cases the whirl velocity is less than the value of u2. The difference between the blade speed at exit and whirl velocity is known as slip.

$$\text{Slip} = u_2 - V_{w2}$$

The slip factor is defined as the ratio between whirl velocity to blade speed,

$$\text{Slip factor} = V_{w2} / u_2$$

Surging Compressor

Let the operation of the compressor at a given instant of time be represented by point $A(p_A, \dot{m}A)$ on the characteristic N_3 curve. If the flow rate through the machine is reduced to $\dot{m}_B$ by closing a valve on the delivery pipe, the static pressure upstream of valve is increased.

This higher pressure, p_B is matched 'with the increased delivery pressure (atB) developed by the compressor. With further throttling of the flow (to $\dot{m}_C$ and $\dot{m}_s$), the increased pressures in the delivery pipe are matched by the compressor delivery pressures at C and S on the characteristic curve.

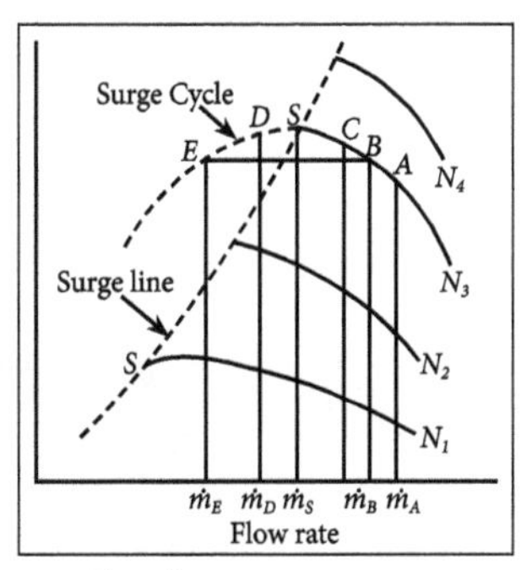

Surging compressor.

The characteristic curve at flow rates below provides lower pressure as are D and E. However, the pipe pressures due to further closure of the valve (point D) will be higher than these. This mismatching between the pipe pressure and compressor delivery pressure can only exist for a very short time. This is because the higher pressure in the pipe will blow the air towards the compressor, thus reversing the flow leading to a complete breakdown of the normal steady flow from the compressor to the pipe.

During this very short period the pressure in the pipe falls and the compressor regains its normal stable operation (say at point B) delivering higher flow rate ($\dot{m}_B$). However, the valve position still corresponds to the flow rate $\dot{m}_D$.

Therefore, the compressor operating conditions return through points C and S to D. Due to the breakdown of the flow through the compressor, the pressure falls further to p_E and the entire phenomenon, i.e., the surge cycle EBCSDE is repeated again and again. The frequency and magnitude of this to-and-fro motion of the air (surging) depend on the relative volumes of the compressor and delivery pipe and the flow rate below $\dot{m}_s$.

Surging of the compressor leads to vibration of the entire machine which can ultimately lead to mechanical failure. Therefore, the operation of compressors on the left of the peak of the performance curve is injurious to the machine and must be avoided.

Surge points (S) on each curve corresponding to different speeds can be located and a surge line drawn as shown in figure. The stable range of operation of the compressor is on the right-hand side of this line. There is also a limit of operation on the extreme right of the characteristics when the mass-flow rate cannot be further increased due to choking. This is obviously a function of the Mach number which itself depends on the fluid velocity and its state.

3.9.2 Stalling on Compressor

As stated earlier, stalling is the separation of flow from the blade surface. At low flow rates (lower axial velocities), the incidence is increased as shown in figure. At large values of the incidence, flow separation occurs on the suction side of the blades which is referred to as positive stalling.

Negative stall is due to the separation of flow occurring on the pressure side of the blade due to large values of negative incidence. However, in a great majority of cases this is not as significant as the positive stall which is the main subject under consideration in this section.

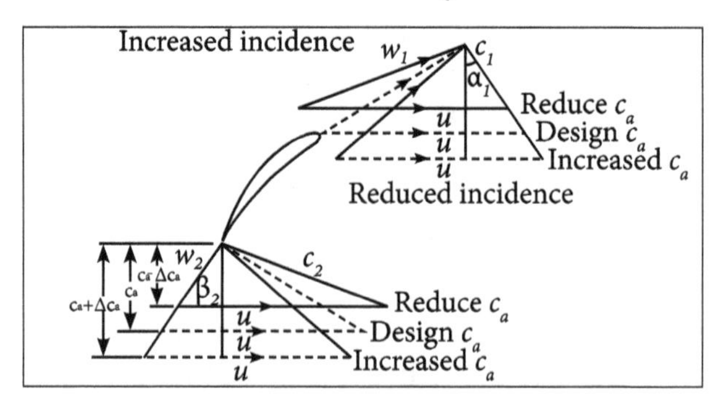

In a high pressure ratio multistage compressor the axial velocity is already relatively small in the higher pressure stages on account of higher densities. In such stages a small deviation from the design point causes the incidence to exceed its stalling value and stall cells first appear near the hub and tip regions.

The size and number of these stall cells or patches increase with the decreasing flow rates. At very low flow rates they grow larger and affect the entire blade height. Large-scale stalling of the blades causes a significant drop in the delivery pressure which can lead to the reversal of flow or surge. The stage efficiency also drops considerably on account of higher losses. The axisymmetric nature of the flow is also destroyed in the compressor annulus.

Problem

1. An axial flow compressor is to be designed to generate a total pressure ratio of 4.0 with an overall isentropic efficiency of 0.85. The inlet and outlet blade angles of the rotor blades are 45 degree & 10 degree respectively and the compressor stage has a degree of reaction of 50 percent. If the blade speed is 220m/s and the work done factor is 0.86, let us determine the number of stages required. And whether the compressor will suffer from shock losses. The ambient air static temperature is 290 K and the air enters the compressor through guide vanes.

Solution:

Given:

Isentropic Efficiency of 0.85,

Total Pressure Ratio of 4.0,

Work Done Factor is 0.86,

Ambient Air Static Temperature is 290 k,

To find:

The Number Of Stages Required.

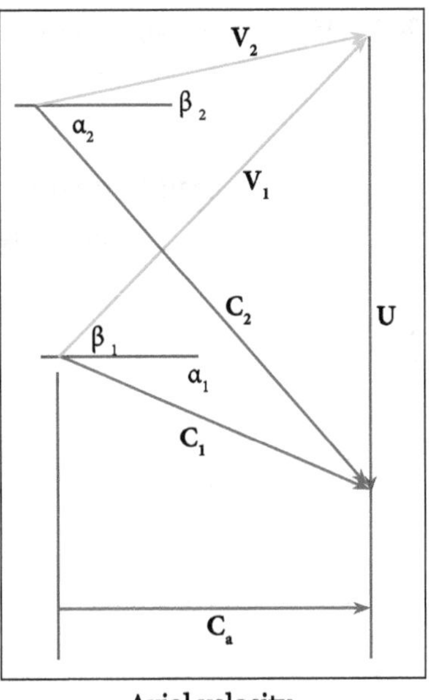

Axial velocity.

$$C_a = \frac{U}{\tan\beta_1 + \tan\beta_2} = 187\,m/s$$

Absolute velocity at inlet,

$$C1 = \frac{C_a}{\cos\alpha_1} = 190\ m/s$$

The per stage temperature rise,

$$\Delta T_{os} = \frac{\lambda \times U \times C_a \times (\tan\beta_1 - \tan\beta_2)}{C_p} = 29k$$

Total temperature at compressor inlet

$$T_{02} = T_2 + \frac{C_1^2}{2Cp} = 331.8\ K$$

Isentropic total temperature at compressor exit,

$$T_{03s} = T_{02} \times \pi_c^{\frac{\gamma-1}{\gamma}} = 493.9K$$

Actual total temperature at compressor exit,

$$T_{03} = T_{02} + \frac{(T_{03s} - T_{02})}{\eta_c} = 522.5\ K$$

Therefore total temperature rise across the compressor,

$$T_{03} - T_{02} = 190.47K$$

$$\text{The number of stages required} = \frac{\text{Overall temperature rise across the compressor}}{\text{Per stage temparature rise}}$$

$$= \frac{190.74}{29} = 6.6 \approx 7$$

Permissions

All chapters in this book are published with permission under the Creative Commons Attribution Share Alike License or equivalent. Every chapter published in this book has been scrutinized by our experts. Their significance has been extensively debated. The topics covered herein carry significant information for a comprehensive understanding. They may even be implemented as practical applications or may be referred to as a beginning point for further studies.

We would like to thank the editorial team for lending their expertise to make the book truly unique. They have played a crucial role in the development of this book. Without their invaluable contributions this book wouldn't have been possible. They have made vital efforts to compile up to date information on the varied aspects of this subject to make this book a valuable addition to the collection of many professionals and students.

This book was conceptualized with the vision of imparting up-to-date and integrated information in this field. To ensure the same, a matchless editorial board was set up. Every individual on the board went through rigorous rounds of assessment to prove their worth. After which they invested a large part of their time researching and compiling the most relevant data for our readers.

The editorial board has been involved in producing this book since its inception. They have spent rigorous hours researching and exploring the diverse topics which have resulted in the successful publishing of this book. They have passed on their knowledge of decades through this book. To expedite this challenging task, the publisher supported the team at every step. A small team of assistant editors was also appointed to further simplify the editing procedure and attain best results for the readers.

Apart from the editorial board, the designing team has also invested a significant amount of their time in understanding the subject and creating the most relevant covers. They scrutinized every image to scout for the most suitable representation of the subject and create an appropriate cover for the book.

The publishing team has been an ardent support to the editorial, designing and production team. Their endless efforts to recruit the best for this project, has resulted in the accomplishment of this book. They are a veteran in the field of academics and their pool of knowledge is as vast as their experience in printing. Their expertise and guidance has proved useful at every step. Their uncompromising quality standards have made this book an exceptional effort. Their encouragement from time to time has been an inspiration for everyone.

The publisher and the editorial board hope that this book will prove to be a valuable piece of knowledge for students, practitioners and scholars across the globe.

Index

Printed in the USA
CPSIA information can be obtained
at www.ICGtesting.com
JSHW051349091023
49903JS00006B/79